Praise for *Wild Ones* by Jon Mooallem

Chosen as one of the Best Books of the Year by
*The New York Times, The New Yorker, The New York Times Book Review,
San Francisco Chronicle, Outside, Scientific American, San Francisco
Magazine, Orion Magazine, Edmonton Journal, National Post* (Canada),
Brain Pickings, NPR *Science Friday, Gizmodo*

"[An] ambitious and fascinating first book . . . [Mooallem] seamlessly blends reportage from the front lines of wildlife conservation with a lively cultural history of animals in America, telling stories of people past and present whose concern for animals makes them act in ways that are sometimes unexpected, sometimes heroic, and occasionally absurd. This is not a book about wilderness; it's a book about us."
The New York Times Book Review

"[Mooallem] does a better job than almost anyone else at taking obscure and often amusing phenomena . . . and spinning tales that ponder their deeper meanings."
—*The New Yorker*

"There is, in short, ridiculously lots to love about Jon Mooallem's *Wild Ones*—starting with its thoughtful and troubling observation that our increasingly extravagant effort at species conservation is a corollary to, as much as a solution for, our habit of rendering wild animals extinct."
—*New York Magazine*

"Intelligent and highly nuanced . . . This book may bring tears to your eyes. If so, they will be drawn out by the tragedy of what we have done and the all-too-often pathetic efforts to turn back the clock. But read through the tears, and you will find yourself more informed, more prepared to make a difference. Mooallem has done those of us who care deeply about nature and wildlife a favor, leaving us justifiably off balance but putting us in a better position to move beyond hubris to pragmatic solutions."
—*San Francisco Chronicle*

"An engaging nature/environment book that goes beyond simple-minded sloganeering."
—*Kirkus Reviews*

"A clear-eyed look at our coy relationship with endangered animals."   —*Nature*

"'If we choose to help [polar bears] survive,' Mooallem writes, 'it will require a kind of narrow, hands-on management—like getting out there and feeding them.' Among a lot of environmentalists, those are fighting words. All respect to Mooallem for having the guts to say them."
—*Outside*

"*Wild Ones* heightens one's awareness of the precipitous position of so many of our animal species, but it's also filled with curiosity and hope. The men and women that Mooallem tails are dreamers, but you wind up rooting for them to keep on dreaming."
—*Smithsonian*

"Mooallem argues conservation is and always has been about fulfilling people's need for nostalgic wildness, however contrived and fictitious it may be. Every generation strives to return the earth to some idealized former state. Although his journey is sobering, Mooallem's conclusion is upbeat: Even small conservation victories matter." —*Discover*

"Mooallem manages to pinpoint something peculiar yet poignant about being human, and as a result, reading his pieces often feels like being tricked by an approachable wink masking a sharp jab to the gut. . . . Be prepared to be surprise-gutted."
—*East Bay Express*

"If I could write this review entirely in smiley faces and majestic animal emojis, I would: *Wild Ones* is easily one of the best books I've come across this year. It's more readable than most novels, stuffed with more fascinating, offbeat trivia than the last three issues of *The New Yorker* combined. . . . It's incredibly well-researched, relevant, challenging stuff."
—*The Portland Mercury*

"This book is dense with both thought and fact. . . . It is written with a vernacularly light touch, shot through with compassion and wit, not to mention open amazement, the only apt response to the story of our monumental hubris."
—*The Daily Beast*

"*Wild Ones* is something altogether different—a deeply curious exploration of America's obsession with wild animals. [Mooallem] has an eye for absurdity and a real fondness for the people whom he writes about. *Wild Ones* makes a compelling case not only for why we care about lost causes, but why we should."
—*The New York Observer*

"A sometimes elegiac, often hopeful, beautifully nuanced look at our relationship with nature and animals headed for extinction . . . [Mooallem] has done his job most excellently. The rest is up to us."
—*SF Weekly*

"This is no emotional animal-rights plea, but a clear-eyed examination of the natural world, its current crisis state, and a bridge between the way creatures are viewed in our culture (cute, cuddly, and disposable) and the way in which they actually exist (at the mercy of progressive environmental disaster). It's also a father's legacy to his child."
—*The Sacramento Bee*

"*Wild Ones* by journalist Jon Mooallem isn't the typical story designed to make us better by making us feel bad, to scare us into behaving. Rather than ready-bake answers, he offers instead directions of thought and signposts for curiosity and, in the process, gently moves us a little bit closer to our better selves. . . . At times poignant, at times playful, at times provocative, *Wild Ones* is altogether fantastic."
—*Brain Pickings*

"A must-read for every nature lover . . . Everyone concerned about conservation should read this beautifully written book."
—*The Buffalo News*

"At its most basic, [*Wild Ones*] is a book about endangered species—but, more accurately, it is a book about the often impenetrable environmental politics and the messy human emotions that surround and frustratingly complicate our interactions with the planet. Mooallem offers what he calls a 'weirdly reassuring'—with an emphasis on both words, as many of his examples are frankly surreal—look at animals, humans, parenthood, and the responsibility of taking care of another living creature."
—*Gizmodo*

"At its most fascinating . . . *Wild Ones* is a natural history, a history of the study of natural history, and an examination of the places wild animals hold in our collective imagination. Mooallem ultimately makes a compelling case that, whatever the outcome, our drive for species conservation is an affirmation of human goodness, and something we should let guide us."
—*High Country News*

"Engrossing . . . Mooallem tells the stories of the dreamers, the misfits, and, yes, the heroes, who have dedicated themselves to defending wildlife." —*Earth Island Journal*

"A gem of a book that opens the doors for some timely collective soul-searching."
—*Toronto Star*

"Easily my favorite nonfiction book of the year, an elegant, dynamic, and even profound look at what we talk about when we talk about nature. I will think about it for years to come." —*Edmonton Journal*

"A poignant tribute to all who try to make the world a better place." —*Booklist*

"It is impossible to express, within the tiny game-park confines . . . how amazing I find this book. I love it line by perfect, carefully crafted line, and I love it for the freshness and intelligent humanity of its ideas. As literary nonfiction, as essay, as reportage, *Wild Ones* is, to my mind, about as good as writing gets."
—Mary Roach, author of *Stiff* and *Gulp*

"I love Jon Mooallem and I love animals, but this book is even better than the sum of its parts. Mooallem makes a persuasive case that wild animals are America's cultural heritage—our Sistine Chapel and our Great Books—and the story he tells is an archetypal American one. Even as the animals are being destroyed by unthinking, unconscious corporate forces, they are also being rescued through the tremendous energy and ingenuity of individuals, men and women who wear whooping-crane costumes, cohabitate with dolphins, and encourage condors to ejaculate on their heads. *Wild Ones* made me proud to be American." —Elif Batuman, author of *The Possessed*

"Part harrowing arctic adventure, part crazy airborne travelogue, and often funny family trek, *Wild Ones* shows us that while saving species might be of debatable value to some, it is maybe in our genes, and definitely in our hearts. Mooallem's analysis of our various environmental movements has the breadth and penetrating clarity of Michael Pollan, but more importantly he makes us wonder even more about a world that is in desperate need of more wonder."
—Robert Sullivan, author of *Rats* and *My American Revolution*

"During the course of his three expeditions, Jon Mooallem collects in the specimen jars of his elegant paragraphs enough ironies, curiosities, insights, and revelations—enough life, wild and otherwise—to stock a mind-altering museum, one unlike any other, in which Martha Stewart has wandered into the polar bear exhibit, and the Hall of North American Animals turns out also to be a hall of mirrors. With Mooallem as your nature guide, you won't look at wild animals—or at *Homo americanus*—quite the same way again." —Donovan Hohn, author of *Moby-Duck*

ABOUT THE AUTHOR

Jon Mooallem has been a contributing writer to *The New York Times Magazine* since 2006 and is a writer at large for *Pop-Up Magazine*, the live magazine in San Francisco. He's also contributed to *This American Life, The New Yorker, Harper's, Wired*, and many other magazines. He and his family live in San Francisco.

WILD ONES

A SOMETIMES DISMAYING,

WEIRDLY REASSURING STORY

ABOUT LOOKING AT PEOPLE

LOOKING AT ANIMALS

IN AMERICA

JON MOOALLEM

PENGUIN BOOKS

PENGUIN BOOKS
Published by the Penguin Group
Penguin Group (USA) LLC
375 Hudson Street
New York, New York 10014

USA | Canada | UK | Ireland | Australia | New Zealand | India | South Africa | China
penguin.com
A Penguin Random House Company

First published in the United States of America by The Penguin Press,
a member of Penguin Group (USA) Inc., 2013
Published in Penguin Books 2014

ISBN 978-1-59420-442-5 (hc.)
ISBN 978-0-14-312537-2 (pbk.)

Book design by Meighan Cavanaugh

146119709

For Wandee

And for Isla, who may not remember any of this

Spock: To hunt a species to extinction is not logical.

Whale scientist: Whoever said the human race was logical?

—*Star Trek IV: The Voyage Home*

CONTENTS

THE WOMAN WHO
COUNTED FISH

My daughter's world, like the world of most American four-year-olds, has overflowed with wild animals since it first came into focus: lionesses, puffins, hippos, bison, sparrows, rabbits, narwhals, and wolves. They are plush and whittled. Knitted, batik, and bean-stuffed. Appliquéd on onesies and embroidered into the ankles of her socks.

I don't remember buying most of them. It feels as if they just appeared—like some Carnival Cruise Lines–esque Ark had docked outside our apartment and this wave of gaudy, grinning tourists came ashore. Before long, they were foraging on the pages of every bedtime story, and my daughter was sleeping in polar bear pajamas under a butterfly mobile with a downy snow owl clutched to her chin. Her comb handle was a fish. Her toothbrush handle was a whale. She cut her first tooth on a rubber giraffe.

Our world is different, zoologically speaking—less straightforward and more grisly. We are living in the eye of a great storm of extinction, on a planet hemorrhaging living things so fast that half of its nine million species could be gone by the end of the century. At my place, the teddy bears and giggling penguins kept coming. But I didn't realize the lengths to which humankind now has to go to keep some semblance of actual wildlife in the world. As our own species has taken over, we've tried to retain space for at least some of the others being pushed aside, shoring up their chances of survival. But the threats against them keep multiplying and escalating. Gradually, America's management of its wild animals has evolved, or maybe devolved, into a surreal kind of performance art.

We train condors not to perch on power lines. We slip plague vaccine to ferrets. We shoot barred owls to make room in the forest for spotted owls. We monitor pygmy rabbits with infrared cameras and military drones. We carry migrating salamanders across busy roads in our palms.

On the Gulf Coast of Alabama every summer, volunteers wait up all night for tiny sea turtle hatchlings to clamber out of their nests in the sand, then direct them safely into the surf, through a trench shoveled down the beach and walled off with tarps to block out the disorienting lights of the condos behind them. And on the Columbia River, at the Washington-Oregon border, the U.S. Army Corps of Engineers lifts endangered salmon out of the water and trucks them around the Bonneville Dam, or barges them through the locks. Sea lions that swim upriver to eat the fish have been hazed with rubber bullets, shipped off to aquariums, and finally just killed; an entire colony of seabirds that had been picking off the juvenile salmon were trapped and relocated; and, at the furthest, most mundane reaches of this almost incomprehensibly sprawling program to protect the fish, the government has even hired ordinary Americans—retirees, housewives, at least one moonlighting concert clarinetist—to work as census takers in a cramped office inside the dam, several stories down, staring through an underwater window to count each and every fish that swims past the glass, an average of 4.5 million fish every year. On the morning I visited, a rail-thin woman named Janet was sitting at an old-fashioned metal desk, six hours into her eight-hour shift, scrunching her eyes with unshakable concentration as fish dribbled by the window one at a time, or swarmed through in rapid-fire mobs. Janet frequently dreams about counting fish, she told me. Once, she sat straight up in bed next to her husband and screamed, "Did you see the size of that one?"

There are many successes. Some aren't as picturesque as we'd like to imagine. American crocodiles, once nearly extinct, have rebounded in Florida largely by colonizing 168 miles of canals dug to cool a nuclear-power plant. And peregrine falcons circle overhead again, thanks in part to ornithologists at Cornell University who, dedicated to collecting new genetic material and reviving the species, put on a specially made leather receptacle they called the "copulation hat" and coaxed captive falcons—one was named Beer Can—to ejaculate on their heads several times a day, every day, for much of the 1970s. Environmentalists are always shouting at America to care more about our planet's many, pressing calamities. But we seem to care deeply enough about our wild animals to strap on the proverbial copulation hat again and again and again.

No one imagined it would come to this. And no one can say how far it will go. In a recent scientific paper, a group led by the government biologist J. Michael Scott conceded that a fundamental belief behind so much conservation—that we can "save" a species by solving a particular problem it faces, then walk away and watch it thrive— is largely a delusion. "Right now," Scott says, "nature is unable to stand on its own." We've entered what some scientists are calling the Anthropocene—a new geologic epoch in which human activity, more than any other force, steers change on the planet. Just as we're now causing the vast majority of extinctions, the vast majority of endangered species will survive only if we keep actively rigging the world around them in their favor. Scott and his colleagues gave those creatures' condition a name: conservation reliance. It means that, from here on out, we will increasingly be forced to cultivate the species we want, in places we protect and police just for them, perpetually rejiggering some asymmetrical balance to keep each one from sliding into extinction. We are gardening the wilderness. The line between conservation and domestication has blurred.

Obviously, there's no hint of all this manual labor in the bedtime stories I've been reading to my daughter. Her animals seem to be getting along fine, in a wilderness that has nothing to do with us. The coyote pups nuzzle. The skunks learn important lessons. *There once was a woman who counted fish* may sound like the start of a simple children's story. But why that actual woman is counting them is too tortuously complicated to ever explain.

THIS BOOK is about finding yourself straddling those two animal worlds—a little kid's and the actual one—and trying to understand both. Or at least it's about me trying to understand them, at first naively and with vague unease, and, eventually, with a mostly futile compulsion to reconcile the two. One of those worlds is real. One is imaginary. But, frankly, for most of us, they both may as well be abstractions.

I grew up in suburban New Jersey, where I loved watching nature shows like *Wild America* but had little to no experience with actual wildlife. (Once, two ducks landed in our backyard and sat in the shade for an entire afternoon. My mother called Animal Control.) And the truth is, as with a lot of adults, my ideas about wild animals probably aren't as different as I'd like to think from the typical toddler's. Once I started looking around, I noticed the same kind of secondhand fauna that surrounds my daughter embellishing the grown-up world, too—not just the conspicuous bald eagle on flag-poles and currency, or the big-cat and raptor names we give sports teams and computer operating systems, but the whale inexplicably breaching in the life-insurance commercial, the glass dolphin dangling from a rearview mirror, the owl sitting on the rump of a wild boar silk-screened on a hipster's tote bag. I spotted wolf after wolf airbrushed on the sides of old vans, and another wolf, painted against a full moon on purple velvet, greeting me over the toilet in a Mexican restaurant bathroom.

Lately, wild animals have become objects of quiet fascination. We now live in a country where it's possible to become an Internet celebrity and get booked on the *Today* show just by posting a YouTube video of an eagle, a fox, and a house cat sitting on your porch doing absolutely nothing. (Last time I checked, Pam Aus's video, titled "An eagle, a fox and my cat all getting along fine on my porch," had more than three million views.) Somehow, it's hard not to be mesmerized by just the idea of them—all these autonomous and unknowable neighbors who just happen to live here, too, on the other side of a fence we only occasionally see over. As the naturalist Henry Beston wrote in 1928, "They are not brethren, they are not underlings; they are other nations, caught with ourselves in the net of life and time."

What I'm saying is, maybe we never outgrow the imaginary animal kingdom of childhood. Maybe it's the one we are trying to save.

I DECIDED TO GO see three endangered species in North America for myself: a bear, a butterfly, and a bird. I figured I'd look closely, ask around, and compare those field notes to the bestiary in my mind.

As it happens, those three animals occupy three different points on a continuum of conservation reliance: over time, humans have manually overridden the machinery of their wildness to different degrees. In the case of the polar bear, conservationists are only starting to ask what we can, and should, do to prop up those animals in a changing climate. With the little-known Lange's metalmark butterfly, we stepped in to resuscitate the species decades ago, hooking ourselves to it like life support and keeping a sometimes ambivalent vigil ever since. And with the whooping crane, we've been on the job even longer, entangling ourselves in the bird's fate more strangely and persistently than anyone could have imagined.

I didn't know any of this at the outset. This isn't a textbook; these three species may not give a fair overview of how modern conservation operates or the range of its accomplishments. I'm not even convinced that their preservation is that important in the bigger, and bleaker, ecological picture. I gravitated to them only because they seemed to have good stories to tell, with a captivating cast of human characters bound up in them. Wherever I went, I found a different crowd of idealists and bystanders encircling the animals—squinting, trying in earnest to see and understand. And, gradually, I found myself squinting at the people squinting at those animals, with the same curiosity. They weren't all government biologists or canonized Jane Goodall types—not even mostly. Instead, I met ranks of impassioned hobbyists, or hobbyists who'd surprised themselves by becoming professionals—many of whom were still trying to justify or refine their own imaginative relationships to those animals and their impulses to help. They were asking the same questions that I was asking as a journalist, and living with the same creeping disquiet about the future that I'd only started to feel as a father.

America has been working out its feelings about wild animals this way since before it was officially America, using them to contemplate its own character. Every country has wildlife, of course, but it's always been different here somehow. When the nation was founded, it didn't have a Sistine Chapel or any Great Books. It had coastlines gushing with oysters and crustaceans, forests crammed with deer and wolves, and out on the frontier, some thirty million buffalo rumbling over the plains as a single, shifting spectacle. Some of the first European travelers to the continent had literally swept up fish with brooms, and letters describing that abundance—how stocked America's pantry was—sped back home as de facto marketing materials to bring over more colonists. In the Old World, wildlife and the right to hunt it were controlled by an uptight aristocracy. But here, as one early traveler boasted, anyone with "strength, sense and health" could gather up enough to live on with minimal effort, no matter how rich or poor. It was a crisp articulation of what we now know as the American Dream.

It never dawned on anyone that those species could be driven extinct. But by the late nineteenth century, it was clear that America was overdrawing its natural wealth, and some took the ongoing extermination as a troubling gauge of where the industrializing nation was heading. The rapid disappearance of wild animals, one early conservationist wrote, was a blight on the "reputation of American citizenship."

From then on, the story of American wildlife would become a story of an infinitely receding Eden. By the 1970s, when the Endangered Species Act made preserving those animals a national priority, our sense of what was at stake enlarged yet again, beyond simple patriotism or even science, tilting toward mushier questions of morality or mysticism.

I found the anxieties and longings of all three of these eras, each roughly a century apart, stitched inside the stories I was following on the ground today. Little by little, they gathered into a kind of cultural history of wild animals in America. I was drawn especially to moments when conservation and popular culture collided, and to historical heroes who were even more idiosyncratic than the folks I was encountering now: Thomas Jefferson, an early American president with a very dorky hobby; William Temple Hornaday, a prideful and paradoxical turn-of-the-century taxidermist; and Joan McIntyre, a Berkeley hippie with a disarming interest in whale sex.

They're not giants of American conservation—they failed quite a lot, in fact. But each seems to have added a new dimension to the meaning of that work, inflecting the ways we've thought and felt about wildlife ever since, and about the consequences of its loss. From the very beginning, America's wild animals have inhabited the terrain of our imagination just as much as they've inhabited the actual land. They are free-roaming Rorschachs, and we are free to spin whatever stories we want about them. The wild animals always have no comment.

I TURNED THIRTY the year my daughter, Isla, was born. I'm part of a generation that seems especially resigned to watching things we encountered in childhood disappear: landline telephones, newspapers, fossil fuels. But leaving your kids a world without wild animals feels like a special tragedy, even if it's hard to rationalize why it should.

The truth is that most of us will never experience the earth's endangered animals as anything more than beautiful ideas. They are figments of our shared imagination, recognizable from TV, but stalking places—places *out there*—to which we have no intention of going. I wondered how that imaginative connection to wildlife might fray or recalibrate as we're forced to take more responsibility for its wildness.

It also occurred to me early on that all three endangered species I was getting to know could be gone by the time Isla is my age. It's possible that, thirty years from now, they'll have receded into the realm of dinosaurs, or the realm of Pokémon, for that matter—fantastical creatures whose names and diets little kids memorize from books. And it's possible, too, I realized, that it might not even make a difference, that there would still be polar bears on footsy pajamas and sea turtle–shaped gummy vitamins—that there could be so much actual destruction without ever meaningfully upsetting the ecosystems in our minds.

That was the most disturbing part somehow—the disconnection. So I decided to bring Isla with me on some of my trips. Even if she didn't remember any of it later, I figured, maybe knowing that she'd seen those actual animals in the wild would make them feel more real. I never predicted, however, that having her with me would change what I saw.

All I wanted was to bring her *out there*. I wanted us to see some wild ones.

PART ONE

BEARS

1.

MARTHA STEWART
ON THE TUNDRA

During the Cold War, a joint U.S.-Canadian military installation was built outside the tiny northern town of Churchill, Manitoba, at the western edge of Hudson Bay. Those stationed at Fort Churchill had several jobs to do, like be ready to repulse the Soviets if they invaded over the North Pole and figure out how to lob nuclear warheads at Moscow through the Aurora Borealis, which was proving, mysteriously, to muck up the guidance systems on their rockets. A lot of the soldiers' time was also spent dealing with a nuisance: hundreds of polar bears that ambled across the tundra there every fall.

In November 1958, for example, one ate a pair of boots at the firing range. Another smashed a building's window, poked his head in, and had to be blasted with a fire extinguisher. At least twenty polar bears were loitering near the mess hall and the dump, and, late one Sunday night, three turned up at the central commissary. Soldiers in station wagons drove them back into the wilderness. One report noted, "The most effective, anti-dawdling weapon has been the small helicopter." Even so, occasionally the bears would rear up on their hind legs and try to tussle with the armored flying machines. One helicopter pilot described how unsettling it was to make a low pass and find "some six feet of indignant polar bear throwing haymakers" with paws the size of dinner plates. After a while, military contractors limited the amount of work done outside at night; the higher-ups decided it would just be easier to stay out of the polar bears' way. "So this is civilization," began one newspaper article about military wives at Fort Churchill.

By the time I arrived, one November a half-century later, the military was gone. The fort had been dismantled and carted off, though two massive, ruined radar domes still sat in the distance like some postapocalyptic Epcot attraction. A dozen specially built vehicles called Tundra Buggies crawled along the network of dirt roads the military had built and abandoned. Each was stuffed with tourists, many of whom had paid several thousand dollars a head to fly to Churchill, now billing itself as "The Polar Bear Capital of the World." They were mostly older vacationers, taken out to the tundra every day to get a glimpse of the animals, then deposited back in town to prowl the gift shops along Churchill's main road, buying polar bear caps and snow hats, polar bear T-shirts, polar bear aprons, polar bear Christmas ornaments, polar bear magnets, polar bear boxer shorts, polar bear light-switch plates, polar bear wind chimes, polar bear baby bibs, and pajamas that say "Bearly Awake."

A Tundra Buggy, if it resembles anything at all, resembles a double-wide school bus propped up on monster-truck tires. Three had pulled off the road to watch a lone polar bear splayed flat at the rim of a frozen pond, asleep in the willows. I was behind them in a scaled-down vehicle known as Buggy One, one of the storied, original rigs of the fleet. Buggy One is now operated by a conservation group, Polar

Bears International. One of the group's videographers was shooting footage of the bear through an open window while the other staff on board tried to sit perfectly still so as not to rattle his tripod. The cameraman had been filming the bear for a long time, in Super HD, hoping it would stand up or do something alluring. Up ahead, tourists filed onto the rear decks of their buggies, training their Telephoto lenses and little point-and-shoots at the animal. It lifted its head once or twice, but that was it. After a couple of minutes, I noticed that the tourists had turned ninety degrees and were photographing us, aboard Buggy One, instead.

It was then that Martha Stewart's helicopter came into view. Everyone turned to watch it as it passed, flying low and very far ahead. Two hundred years ago, Arctic explorers described polar bears leaping out of the water and into boats, trying to "resolutely seize and devour" whichever dog or human being was sitting closest to their jaws, unprovoked and absolutely undeterred even if you tried to set the bear on fire. Now Martha Stewart had come to Churchill to shoot a special segment about the bears for her daytime television show on the Hallmark Channel.

Polar Bears International had been working in a loose partnership with Martha Stewart for many months in advance to handle logistics for her shoot. The group was trying to ensure that Martha told the right story about the animals. It isn't enough anymore to gush about how magnificent or cute polar bears are, as the many travel writers and television personalities that came to Churchill over the years had tended to. The stakes were too high now—too urgent. Climate change had put the bear in severe jeopardy. According to a 2007 study by U.S. government scientists, two-thirds of the world's polar bears are likely to be gone by the middle of this century. And, of course, that's only one of many dispiriting prognoses trickling into the news these days. Another study predicts that climate change may wipe out one of every ten plant and animal species on the planet during that same time. Another claims seven of every ten could be gone. Tropical birds, butterflies, flying squirrels, coral reefs, koalas—the new reality will rip away at all of them, and more. The projections range from bona fide tragedies to more niggling but genuinely disruptive bummers: tens of millions of people in Bangladesh are likely to be displaced by sea-level rise and flooding; the Forest Service warns of maple syrup shortages in America.

The polar bear, in other words, is an early indicator of all this other turmoil coming our way. It is, as everyone on Buggy One kept telling me, a "canary in the coal mine"—that was the phrase they used, always, with unrelenting discipline. The animal had become a symbol for some otherwise inexpressible pang—of guilt, of panic—that can burble into the back of your mind, or the pit of your stomach, when you think about the future of life on Earth. But, Polar Bears International was arguing, it could also be a mascot—a rallying point. By now, bears are all but guaranteed to disappear from a lot of their range. But the science suggests that there's still time to slow climate change down and, in the long term, keep the species—and many others along with it—from vanishing entirely.

Practically speaking, this leaves conservationists like Polar Bears International in a unique and sometimes disorienting position. Unlike with other species, the central threat to polar bears isn't something that can be tackled or solved on the ground, out in the immediate ecosystem. The only meaningful way to save the polar bear now is to influence the energy policies and behavior of people who live thousands of miles away—which means, in part, influencing influential media personalities like Martha Stewart. At some point, polar bear conservation stopped being solely the work of scientists and became the work of lawyers, lobbyists, and celebrities as well. The bear is dependent on the stories we tell about it.

After spending the fall in Churchill, Polar Bears International's president, Robert Buchanan, would head back home to the United States and start traveling from city to city, hosting talks by scientists and zookeepers, trying to use the appeal of this one charismatic animal to inspire people to reduce their own carbon footprints, however slightly—to drive less, to buy recycled goods. In Kansas City, PBI had partnered with the hardware chain Lowe's to get inner-city kids to weatherize their neighbors' homes, saving energy for heating and cooling. In suburban Connecticut, they'd cosponsored "Polar Bear Empathy Day," at which members of the local Polar Bear Club, in a reversal of their traditional cold-water swims, put on heavy parkas and stood on a scorching beach in July to show solidarity with the bears in an overheating Arctic. All together, Robert regarded these strategies and stunts as a kind of psychological guerrilla warfare. "Polar bears are in serious friggin' trouble," he told me that morning on Buggy One. "But until you change the consumer's attitude, you're not going to change the policy or the political will." By "consumer," he presumably meant "citizen."

It was a marketing gambit, after all. And Robert, a big man who talks in a languorous growl, felt very comfortable relating to it in those terms. This was his retirement. In his thirty-five-year career, he'd risen to marketing director at Joseph E. Seagram & Sons, overseeing beverage and alcohol brands during the heyday of the corporation, when it owned Universal Studios and a large share of the music industry and was producing flashy wine-cooler commercials starring a young Bruce Willis. Robert handled cognacs and whiskey and Tropicana orange juice. "I take products to market," he said. "I'm a marketer." Now he'd put himself on the polar bear account.

Robert was literally trying to control the image of the polar bear in Churchill before that image was broadcast around the world. Churchill turns out to be the best, most convenient place in the world to see or film polar bears in the wild. (When you see a wild polar bear on TV or the Internet, the chances are good that you're looking at a Churchill bear.) Because Polar Bears International operates in close partnership with a tour company in Churchill that owns the majority of the permits and vehicles needed to access the animals on the tundra, the group has been able to intercept most of the major media that come through town. They install biologists and climatologists on the reporters' buggies like scientific press agents, trying to make sure an accurate narrative comes across, and they provide B-roll footage of bears plunging into melting slush to help newscasters illustrate the problem. In past years, though, PBI had gone out of its way to help television crews only to feel betrayed by the finished product: the reporters ignore climate change altogether, or regurgitate the junk theories of climate change deniers. Most television crews are now asked to sign memorandums of understanding, outlining certain guidelines, before working with PBI. (As a rule, one PBI staffer told me, Robert regards all journalists as "pirates and thieves.") But that fall, Martha Stewart hadn't signed one. And in the days before her arrival, her producers had become a little incommunicative about their plans. We'd all headed out on Buggy One that morning because PBI had hoped to tour the tundra alongside Martha and her crew, docking back-to-back with Martha's buggy to pass people back and forth periodically for interviews. But that was starting to feel unlikely now. Though no one was quite saying it, there seemed to be concern that Martha Stewart was going rogue.

Everyone on Buggy One sat around for quite a while, not speaking much, while the cameraman stayed locked optimistically on the bear lumped in the mud outside, trying to gather whatever stock footage he could. Eventually, the polar bear got up and walked away. The cameraman shrugged.

We drove on. We looked for more bears. People noodled on their laptops and iPads. It felt aimless. Then, sometime after lunch, a voice crackled over the radio. It sounded like we would finally rendezvous with the Martha Stewart people and do a little filming on Buggy One. A PBI employee started wiping down the vehicle's counters with wet wipes. She hung everyone's parkas on coat hooks in the back. But when we got closer, we saw Martha's buggy receding in our windshield, fairly rapidly. "Oh, they're moving now," our driver groaned.

All day, a strange paparazzi-like triangle had been materializing: Martha wanted good access to polar bears, and PBI wanted good access to Martha. I wanted to watch the whole process of brokering access, since I was quickly understanding that the media relations dimension of polar bear conservation was a critical part—maybe the most critical part—of the preservation of this five-million-year-old species.

Our driver pushed Buggy One as fast as it would go, which wasn't very fast, trying to gain ground. Something that I'd kind of suspected for hours was suddenly obvious: we were chasing Martha Stewart across the tundra.

ALL OF THIS WAS HAPPENING at the end of 2010, when it was getting easy to wonder how potent a symbol the polar bear even was anymore. A few years earlier, the bear had helped install climate change as the central issue of American environmentalism. But since then the percentage of Americans who even believe the issue is real had plunged. A poll released a few days before I'd arrived in Churchill, from the Pew Research Center, showed that only 34 percent of Americans believed that humans were altering the climate. Congressmen were once again openly dismissing the idea as comical, while thrashing bills to limit greenhouse gases. Back home in San Francisco, when I took my daughter, Isla, to the California Academy of Sciences one day, I noticed that the museum was scrapping its exhibit about disappearing glaciers and polar bears. The exhibit had proved unpopular and was mostly ignored, and was being replaced with an exhibit about the worsening droughts, forest fires, and floods that climate change would bring to California. The new message, a museum director told me, was "Never mind the polar bears. Concentrate on how bad it's going to be for you."

It was starting to feel as if the polar bear had jumped the shark. Part of the problem was that the symbolism of the bear had become so ingrained, so legible, that it was getting overexposed. Polar bears were everywhere now, receding back into the pop cultural noise that environmentalists had originally used them to cut through. There was "Floe the Polar Bear," the mascot for a natural gas company, and the "Time to Care" limited-edition gunmetal polar bear wristwatch. A television commercial introducing Nissan's plug-in electric car, the Leaf, showed a polar bear being forced off his shrinking ice floe and walking sullenly through forests, along busy highways, and all the way into an unnamed city, where he found a commuter getting into a Leaf and gave the man a long, almost pitiable hug. The hug goes on so long that you start to realize the bear is doing more than just saying thank you—he's absolving the man, and maybe also just needs to lean on the guy out of sheer exhaustion.

The predicament of actual polar bears, meanwhile, seemed only to be getting worse. Projections that sounded like sad science fiction scenarios when laid out years ago in scientific papers were being observed around Churchill more frequently. Polar bear cannibalism, for example. The previous autumn, a hungry male bear was seen killing and eating a polar bear cub after separating it from its mother. I saw a photo of the cub's severed head hung sideways from a long bolt of fur in the larger bear's mouth; the male faced away, but the cub's dead face looked straight into the camera, its fur stained salmon pink with blood and snow. This was one of several photos of the cannibalism incident floating around. The whole thing happened in front of a Tundra Buggy full of tourists.

Biologists also believe that more cubs will be abandoned around Churchill in the coming years, as mothers either become too malnourished to nurse their young and leave them, or simply drop dead of starvation. And so the Assiniboine Park Zoo in Winnipeg was putting the finishing touches on a new polar bear "transition center"—the province of Manitoba wanted to establish a place where those cubs could be flown, triaged, and fattened back up, before being distributed to zoos. It was basically an orphanage, but with an adjoining classroom to accommodate school field trips. The orphaned cubs would be leveraged as an educational tool—stirring proof of what climate change was actually, already doing. The trick would be to explain it to the kids in a way that instilled hope, and didn't just terrify them. "I think the message is going to have to be pretty carefully crafted," a zoo official told me.

The flamboyant crusade for the polar bear underscores a quieter truth. Wildlife managers sometimes talk about a species' "cultural carrying capacity," meaning that it's not just the availability of food or habitat that determines how well a species will fare, but also—if not mostly—our willingness to tolerate it or help it along. Normally, it's the cultural carrying capacity of animals totally unlike the polar bear that seems to matter most in their survival—animals that live in close proximity to us like deer or wild turkeys. But by now, with humans exerting such overwhelming influence on the planet, it seems that even the welfare of a far-flung creature like the polar bear depends on our goodwill. Here, at the lip of Hudson Bay, was an animal whose cultural carrying capacity had suddenly become tremendous; that, in fact, had reached into the deep murk of human psychology and re-formed its entire reputation, from vicious monster to sweet and cuddly star. Still, its future wasn't so bright. Partly, I'd find, that's because the threat of climate change is just so colossal and complicated. And partly it's because you can get so mesmerized and impressed by the polar bear's charisma that the truth of its predicament gets lost.

The longer I stayed in Churchill, the more stories I found mush-rooming out of this one grungy town: stories about Thomas Jefferson, teddy bears, opossums, and sharks, all showing the convoluted and sometimes arbitrary ways in which we forge feelings about wild ani-mals; and also the story of the Endangered Species Act, the law through which we impose those feelings on the landscape, shaping it so that certain off-brand animals are left to fend for themselves while the icons we love are battened in place.

In the twenty-first century, how species survive, or go to die, may have more to do with Barnum than with Darwin. Emotion matters. Imagination matters. The way we see a species can impact its standing on the planet more than anything covered in ecology textbooks. And so, suddenly, the question on the tundra was, What did we all think that we were seeing when we looked at these bears?

WHEN BUGGY ONE FINALLY caught up to her, Martha Stewart's Tundra Buggy—Buggy Two—was parked. Her crew was filming an imposing, solitary male bear, standing on its hind legs behind a snow-drift. It was a lucky find. No one likes to shoot polar bears without snow—it just looks wrong—and it had been so warm lately that there was little snow in the viewing area, just rocks, slush, and dust. But here the wind had aggregated the last evidence of a snowfall into a curling ridge, several feet high. The bear began pawing the top of it. Then it climbed under the curl of snow and stretched its paws and legs straight out, like a drowsy cartoon grizzly settling into a hollow log. After that the bear got out and bashed the snowdrift to the ground. It was a stellar action sequence. "Martha's going to be happy," a woman on PBI's staff whispered.

Our driver backed up Buggy One to dock with the rear of Martha's vehicle. Quickly, an editorial meeting started taking shape on the two conjoined decks. One of Martha's producers, a man with a biting, hard-to-place accent, explained the shot he wanted: Martha would watch the sunset and chat with Dr. Steven Amstrup, who had just retired as one of the United States government's top polar bear biologists to join PBI as its senior scientist. The two of them would stand on the back of Buggy One while the crew filmed them from the other vehicle, parked far enough away to get a wide shot.

The talent boarded Buggy One. The driver pulled out and got into position. "Talk about the tundra itself! The ecosystem! What the tundra comprises! How it's not just rocks and snow!" the producer yelled from the other deck. But, much to his mystification, the Polar Bears International employee driving Buggy One had ground the vehicle back into gear and was now executing a series of harried, multi-pointed turns. (The driver was worried he'd parked in such a way that the PBI logo on the back of Buggy One wouldn't make it into the shot.) The producer seemed to be in a hurry. The sun was slipping away. He kept shouting his directions: "The *actual* tundra! What does it comprise?"

By now, the bear near the snowdrift had wandered off. A few of us had watched it issue a cascade of liquid crap, then turn to smell it. When the cameras finally rolled and Martha and Amstrup strolled onto the deck of Buggy One and hit their marks, I noticed another polar bear, or maybe the same one from before, loping off, out of frame, away from the commotion and back into the nothingness where it somehow makes its living.

"Now, polar bears," Martha Stewart said to the biologist, beginning a take. "Where do they come from?"

2.

AMERICAN INCOGNITUM

The first polar bear tourists started arriving in Churchill thirty years ago to see a bloodthirsty man-killer—an enigmatic Lord of the Arctic. Now we'd all come to see a delicate, drowning victim. It's a baffling shift that I found written into the history of the town.

Polar bears are generally not cooperative tourist attractions. They are elusive and solitary—it's part of their allure—each a nomadic white blip, wandering thousands of miles of the Arctic independently, searching for prey at the fringes of human habitability. But Churchill's roughly 950 bears constitute one of the southernmost polar bear populations in the world. They have been forced to evolve an atypical lifestyle in order to stick it out at the tip of their species' range. Like all polar bears, they spend the winter and spring on sea ice, hunting seals. (When seals poke their heads out of the ice to breathe, the bears yank them up and peel them open like a banana. They eat only the fat.) But the ice on Hudson Bay melts every summer—the bay becomes open water. Though the bears stay on the ice as long as they can, they are eventually forced to decamp. They spend the next several months marooned on land, unable to hunt seals and living off reserves in a kind of walking hibernation. Then, in the fall, they gather near the coast outside Churchill, where the ice is usually first to re-form, anticipating freeze-up and their earliest opportunity to get back out and hunt again. They have been migrating this way since long before humans decided to build a town there, herding themselves together every fall to wait.

So much about modern-day Churchill is explained by this quirk of ecology. It's the only place where such a large-scale polar bear tourism industry is logistically possible—where so many polar bears not only congregate, but congregate predictably in the same place, at the same time every year. Also, they happen to congregate outside an actual town, with an airport, a train station, and a grocery store. Churchill, in short, is that rare place where large numbers of people and polar bears can comfortably commingle. And after the military pulled out, taking the town's economic reason for being with it, it didn't take locals long to figure out what this meant. It meant that Churchill's polar bears, more easily than any other polar bears on Earth, could be monetized.

The military was gone by the early seventies. Churchill's population almost instantly dropped from five or six thousand people to about sixteen hundred. Fort Churchill had been a bustling, modern community, with a youth hockey league, and lobster flown in for New Year's Eve. But the actual town of Churchill, home to the civilians who stayed behind, didn't even have phones or sewers. In 1968, a government report described the "unparalleled squalor" of the town; in one year alone, more than twenty people were said to have either frozen to death in their homes or died in fires. Outdoor sandboxes served as toilets, and raw sewage ran through the streets during the spring thaw.

The industriousness of those who stayed after the military pullout was impressive. There's always been a committed culture of scavenging in Churchill. (Because no roads connect Churchill to the rest of civilization, shipping goods in, and waste out, is expensive.) Now the fort's abandoned buildings were gradually dismantled and repurposed. I met one couple who ripped apart the military's jail and reclaimed the wood to build a new house. The floor of the old bowling alley was pried up and installed in what was now the lobby of my hotel, and an old Nike missile adapter sat in the corner, holding a ficus. The military had left an ambulance, but there was no one to drive it. So the woman who drove the town's taxi took that on, too.

The military, locals told me, also left a tremendous amount of scrap: school buses, bombardiers, airport-grade snow blowers, fire trucks. It wasn't uncommon for young guys with nothing to do to head to the scrap dump with a can of gas, a car battery, and some tools; in no time, they could get something up and running just for the pleasure of crashing it into something else. A garage owner named Len Smith started cobbling together primitive all-terrain vehicles and heading into the emptiness east of town. "He just liked it," a friend of Smith's told me. "He liked being out there." There were no more military sentries shooting bears that came too close to the fort. There was no more fort. All the noise and lights were gone. The human presence had been winnowed down to an almost negligible little settlement. All kinds of wildlife were beginning to creep back into the area and rebound.

SMITH HIRED HIMSELF out to photographers who wanted to take pictures of polar bears at Cape Churchill, a crook of land about thirty miles east of town. Among them was an American photographer named Dan Guravich. The two men would camp on one of Smith's vehicles for two weeks at a time, hefting balls of lard over the side to lure in, or position, the bears. (Feeding bears wasn't explicitly illegal then.) In 1978, Guravich published a ten-page spread of his bear photos in *Smithsonian* magazine, attracting even more photographers to the area.

Until then, the average person's only exposure to polar bears would have been at zoos or the circus. (The year Guravich's photos ran in *Smithsonian*, a tiny East German woman in a fur-lined miniskirt named Ursula Böttcher was touring America with her troupe of ten performing polar bears as part of the Ringling Bros. and Barnum & Bailey Circus, closing her performances by shouting, "Once again, they don't eat me!") The pictures emerging from Churchill in the late seventies were, at the time, some of the best-ever shots of the animals in the wild, and the first many Americans saw. It was a simple issue of access: Before Churchill, photographing wild polar bears meant mounting an expedition into the infrastructureless white wasteland of the rest of the species' range. But here the land brimmed with the animals every fall, like clockwork. One of the first scientists to study polar bears in Churchill, the Canadian government biologist Ian Stirling, has described his astonishment at sighting 239 bears over the course of four hours in 1970. At first, he mistook three dozen males staggering around an area the size of a football field for a herd of caribou. For anyone wanting access to wild polar bears—photographers, scientists—Churchill was an unimaginable El Dorado.

In 1980, Len Smith got a call from a National Geographic Society crew that wanted to shoot a television special about the town and its polar bears. Smith built a new tanklike vehicle to guide the men around in. He called it the Tundra Bus, but would settle eventually on Tundra Buggy. Before the film crew arrived, he invited friends in town on a celebratory test run. Folks were having a great time, eating hors d'oeuvres and drinking champagne, when, a few miles east of town, they heard a foreboding thunk and saw one of the buggy's wheels roll freely past the vehicle.

National Geographic's film aired on public television in early 1982. It set ratings records, putting Churchill and its bears on the map. When I asked people around town about the history of tourism in Churchill, they usually started their story with that broadcast. The show was called *Polar Bear Alert* and opened on a young couple, walking through town with a stroller. The man slung a rifle over his shoulder while the narrator, the actor Jason Robards, described Churchill as the "one place in the world where the great white bears roam the streets, dangerously immune to the presence of their only enemy: man." A good deal of time was spent rehashing bear attacks in town. A man who lost an arm was interviewed, and a mother in pajamas was shown running out of her front door in the dark, waving a shotgun and screaming: a bear had approached her house in the middle of the night and had to be shot by a wildlife officer. For the film's most memorable scene, one of the show's producers, James Lipscomb, was placed inside a rebar cage with a camera and left on the tundra. "No one knows how the bears will react to a man in a cage," the narrator says. Next comes a harrowing point-of-view sequence of polar bears rattling, pummeling, and chewing the thing.

"I sort of lost respect for the Geographic after doing that with them," a local named Paul Ratson, who worked bear security for the crew, told me one afternoon. Ratson is fifty-five, a hefty, forthcoming guy in a tweed cap who moves around town with multiple walkie-talkies crammed in the waistline of his overalls. He runs one of the smallest tour companies in town, Nature 1st, and is constantly clicking into one walkie-talkie or another to reschedule an airport pickup or redirect a school bus. (The entire tourism industry relies on a fleet of old school buses, some of which are said to date to the military days. Given the expense of shipping new vehicles in, they are perpetually jury-rigged to last another season.)

Ratson, like a lot of people in town I met, resented how sensationalized *Polar Bear Alert* turned out. The town sees itself as coexisting with its polar bears, with caution and respect. Ratson felt the show painted people in Churchill as being hunkered down in fear. And, worse, it made the bears look like monsters, twisting the animals to fit the stereotype that viewers would already know. "The guy in the cage, he's going, 'Oh no, the bear really wants to eat me!'" Ratson said. "But these bears are bored stupid. You put a guy in a cage and it smells like food, what's the bear going to do? A lot of film crews want to portray the bear as this reckless, violent, bloodthirsty man-eating creature," he told me, grimacing. "It's like the way they do sharks."

In fact, it's exactly how they do sharks. Before filming *Polar Bear Alert*, James Lipscomb had codirected a feature documentary called *Blue Water, White Death*. That film chronicled an obsessive six-month expedition to South Africa, Sri Lanka, Madagascar, and Australia led by the department-store heir and serial adventurer Peter Gimbel, to film some of the first footage of great white sharks in the wild. Gimbel's idea was to follow commercial whaling vessels around. After a kill, as the vessels tied up the whale and waited for the larger ship that would process it, Gimbel lowered his cameramen into the water in specially built cages to film the sharks ripping bits from the carcass. *Blue Water, White Death* begins with plumes of blood slowly expanding in water, and in its violent finale, a great white batters one of the cages, denting and warping the bars as it thrashes. The film was a hit—a *New York Times* reviewer called the action "as poetic as anything I've seen on the screen in a long, long time." It inspired the author Peter Benchley to write *Jaws*.

By re-creating the cage sequence with polar bears, Lipscomb was one-upping himself—getting as close as possible to a dangerous animal in order to deliver the most innovative and thrilling footage. (*Blue Water, White Death* and *Polar Bear Alert* are now looked back on as cutting-edge films, and the cameraman-in-the-cage has become a well-worn stunt in wildlife filmmaking. In 2008, for example, Animal Planet broadcast a three-part series in which a male model got in a Plexiglas cube amid grizzlies, lions, and crocodiles.) The effect of *Polar Bear Alert* was twofold: it announced to the world that there was a relatively accessible place to see these vicious animals, and it reinforced that one-dimensional, terrifying image of them. And, as it happens, Churchill's bears quickly made good on that reputation.

THE YEAR AFTER *Polar Bear Alert* was broadcast, freeze-up of Hudson Bay was extremely late. The bears were hungry, and, still unable to hunt seals, they were emboldened to lurch into town. One night in late November, shortly after midnight, a down-on-his-luck local character named Tommy Mutanen slipped into the burned-down remains of the Churchill Motel. The motel had sat there for days after a fire, untouched, while the town waited for a fire inspector to fly up and determine it wasn't arson. ("Who cares if it was arson?" one man told me. "That place needed to go.") Mutanen made for the motel kitchen, looking for something worth scavenging. Everything inside the meat locker had been protected from the blaze. So he stuffed his parka's pockets full of raw meat and skedaddled.

"Old Tom Mutanen walked with a gimp leg and a crutch," a man who'd been a pallbearer at Mutanen's funeral told me. There was nothing resembling a chase. Either the polar bear was waiting for Mutanen outside the motel, or it had been in the kitchen alongside him the whole time, biding its time in the dark. The bear seized Mutanen by his skull, and dragged him over a snowbank to the doorstep of the shop across the street. All the bars in town were just letting out, and a crowd of people began yelling at the bear, throwing ice and tools at it—whatever they could find. This only antagonized the animal; it was fixed on its meal. Finally, a young tour guide named Mike Reimer came sprinting barefoot from his apartment with a rifle. He shot the bear right where it stood, on top of Mutanen's broken frame.

A young television reporter had arrived in Churchill that night to do a story on the hotel fire. After her piece on the Mutanen incident aired, more reporters descended on Churchill. Six days earlier, a man had been photographing a rare ivory gull out the window of a Tundra Buggy when a bear, which had been sitting under the chassis, sprang to its hind legs and locked its mouth around his arm, tearing it open with two bites. The press stories about the Mutanen attack mentioned this incident, too. The two together gave the impression of a trend: a coordinated spree.

There was concern that the Mutanen mauling, the town's first lethal bear attack since the sixties, would scare away tourists and crush the industry's new momentum, leaving Churchill to founder again without any obvious economic opportunities. But the news had just the opposite effect. The danger was the attraction. As one veteran tour operator told me, "If there was a chance that you could be ripped to pieces by a polar bear, people were into it. If anything, more people started coming." Two hundred and fifty-four tourists were booked on a charter from New York the following year.

"WE COMPETE IN THE LONG-HAUL, high-yield market," John Gunter told me one afternoon. Gunter is in his early thirties and slight, with a styled wave of hair that pours over his forehead. He lives in Winnipeg, but has been spending every fall in Churchill running Frontiers North Adventures, the company his parents founded in the nineties. It's now the largest and most visible outfit in the town's roughly $7-million-a-year tourism industry. We were eating a couple of bowls of minestrone soup, in a restaurant the Gunter family recently bought half of.

The Gunters first came to Churchill in the eighties, when John's father was sent to manage the Royal Canadian Bank branch in town. At that time, the tourism industry consisted of a few hardscrabble one-man operations—locals who, like Len Smith, hauled people out to the tundra on vehicles they built themselves, operated little hotels, or rented rooms in their homes. Hard-core wildlife photographers were willing to go through the trouble of cobbling together travel arrangements from all these entrepreneurs, individually, by mail. The Gunters' innovation was to package these proprietors together and market all-inclusive vacations. They made the process of booking a trip to Churchill more like booking a stay at a Caribbean resort, and eventually bought the Tundra Buggy business from Len Smith when Smith retired to Florida. They began advertising through mainstream outlets like automobile clubs and the AARP. They were opening Churchill's frontier to a wider spectrum of tourists—namely, Gunter told me, what the industry calls "classic tourists."

I'd been in Churchill for a couple of days at that point. The gaggles of classic tourists were impossible to miss. They were older—the average age of travelers to Churchill is sixty-one, according to one study—and moved between scheduled activities in chaperoned cohorts. The classic tourists seemed to delight in everything, wherever they went. It was always uplifting to hear the swishing of their snow-proof pants as they rose smiling, en masse, from their breakfast table at Gypsy's Bakery and filed onto a school bus waiting outside, to light out for their next low-impact adventure.

Over the years, the motivations of Churchill's tourists seem to have changed just as dramatically as their demographics. The first wave of visitors to Churchill were coming to photograph the bears, but tended to have a broad appreciation for the landscape and its ecology. By comparison, locals told me, the current crop feels more myopically bear-centric—they come to Churchill to tick something off a list. A 2010 study in the journal *Current Issues in Tourism* on the phenomenon of "Last Chance Tourism" found that a majority of people traveling to Churchill were coming because they wanted to see polar bears before they went extinct. As one industry veteran put it, "People are coming to see the last dinosaur now."

CHURCHILL'S POLAR BEARS are almost certain to be the first polar bears wiped out by climate change. This is a corollary of the same unconventional migration that strands them here every fall to be marveled at by tourists. Polar bears in Churchill are already pushing the limits of what's possible for their species. Surviving in a place where the primary habitat—ice—simply vanishes for a few months each year has meant evolving one of the longest fasting periods of any animal on Earth. The bears' lifestyle is almost literally feast or famine: bulking up all winter, then living off that accumulated energy while waiting out the summer on land. It means flirting with a breaking point at the lowest point of that cycle every year, just before the ice re-forms. (A healthy female bear might lose two-thirds of her body weight, only to put it back on after freeze-up.) And it leaves very little wiggle room to adapt if conditions change.

A body of scientific research, including work by a University of Alberta biologist, Andrew Derocher, has shown that the ice on Hudson Bay is now breaking up three to four weeks earlier than it did in the 1980s, pushing the bears onto land earlier in the summer, and for a longer stretch of time. In short, they are being given less time to eat, then forced to fast for longer. Consequently, in the last twenty years, Churchill's bear population has declined by more than 20 percent.

Polar Bears International had flown Derocher to Churchill while I was there. Among other things, they wanted him on hand for the Martha Stewart taping. He is a towering, affable guy, and among the world's most respected authorities on polar bears. When I knocked on the door of the sparsely furnished rental house PBI had put him up in, he and one of his graduate students had just finished watching the new Nissan Leaf commercial online—the one with the polar bear hugging the commuter. "It really is kind of depressing that the animal I've spent a long time studying is being used as the motivator to buy a car," Derocher said. He was ambivalent about the way polar bears had infiltrated popular culture in general—"I can't go anywhere without seeing polar bears," he told me—and was skeptical of electric cars as a solution to climate change in the first place, since they'd wind up plugged into an electric grid that often relied on coal.

Derocher has studied Churchill's bears since he was a student in the mid-eighties. He explained that the bears lose about two pounds every day that they're on land. And so a bear's body mass when it comes off the ice in the spring is one measure of how likely it is to survive until the next freeze-up, of "how much gas is in the tank." Derocher's mentor, Ian Stirling, has described healthy bears as resembling "big tubs of jelly with little, stubby legs sticking out." By fall, a bear in poorer condition will have a leaner profile, like a hunting dog. A starving bear will start to resemble a duffel bag with a bunch of wrenches clunking around inside.

Still, the big drain on the population so far doesn't appear to be adult bears dying of starvation, but a flickering out of the new generations meant to take their place. Derocher has shown that females with less fat on them when they come off the ice in the spring give birth to smaller cubs. And smaller cubs are less likely to survive. Slimmer mothers also produce fewer cubs. (Twins and triplets in the population are becoming less common.) Females that become so weakened while waiting for freeze-up may also shut down their ability to nurse to conserve energy, and their cubs will starve. Or else they forgo reproduction in the first place. Derocher has shown that females whose body mass fell below about 415 pounds were unable to reproduce successfully, and they are all being quickly driven in that direction. (Derocher tracks females with satellite collars; he recently had to order new, smaller collars, because the old ones are now too big for the shrinking bears.) If all the females reach that reproductive point of no return, with no young surviving, it would just be a matter of waiting for the adults bears to die out. Suddenly, the polar bears of Churchill would be a society with no children.

The general trend, from here on out, will be for increasingly longer summer seasons. The bears will return to the ice later and later in the fall, in worse and worse condition, and with fewer and fewer cubs. The coastline where the Tundra Buggies roam may gradually morph into something akin to a polar bear refugee camp, a holding area for despondent, gaunt animals with no ice in front of them and no place to go. There will be good years among the bad years—colder, with earlier freeze-ups. But there may also be catastrophic years, when the ice melts extremely early in the spring and re-forms extremely late in the fall or winter, and a disproportionate chunk of the population gets knocked off all at once, unable to wait it out. Nothing says that Churchill's polar bears will disappear in a gradual, linear way, Derocher told me. It could be quick if just the wrong succession of sudden shocks plays out. Recently, at a meeting of Canadian wildlife officials, Derocher suggested that plans ought to be drawn up for short-term crisis scenarios: "What if some sort of freak thing happened, and freeze-up didn't start until late December? What are you going to do if you end up with forty bears that are starving? Do you want to euthanize them? Do you want to pick them up in a helicopter and release them somewhere else? Do you want to feed them? Bring them to a zoo?" His point was acknowledged but not dwelled on, he said.

By 2050, Derocher told me, Hudson Bay is expected to stop freezing at all; it will be open water year-round. No matter how the particulars in Churchill have played out by then, the sea ice—the polar bear's true home—will have vanished entirely. If any bears are left outside town, they will vanish with it.

There are eighteen other populations of polar bears on the planet, and over the next century, a different story will unfold in each. Seven of them are already known to be declining, though less steeply than Churchill's. Some of the populations at the highest latitudes may increase temporarily, as more seals and walruses—animals the bears eat—shift their own ranges north, chasing a colder climate. Some bear populations might only *appear* to increase, since polar bears will be hungry enough to roam closer to Arctic villages in search of food, and people will see the animals more frequently. And a small number of bears, scientists project, will survive into the next century on pockets of ice around northern Greenland and the Canadian Arctic Islands— and maybe indefinitely if, in the near future, the world actually starts to make the sort of wholesale, politically unpopular shifts needed to slow climate change.

That is, when environmentalists like Robert Buchanan say they are fighting to save the polar bear, what they mean is bleaker than it sounds: they are fighting to save *some* polar bears, these ones in the very far north that still, theoretically at least, could be preserved like a keepsake. Robert, for his part, seems to prefer to talk more about humanity's opportunity to preserve the polar bear *species,* rather than break down the scenarios for each individual population. He never ducks the scientific reality. But it would undermine his message of building hope and motivating action to concede, while preaching to television crews on Buggy One, that it's too late to save the particular polar bears wandering behind him in the shot. In even a best-case scenario, the species will be withdrawn into those shrunken sanctuaries at the top of the world and recede into an even more distant and wild abstraction, a creature once again unreachable by tourists and celebrities and their cameras. At that point, the bear will live on mostly in whatever legends we are beginning to tell about it now.

ONE NIGHT, I was invited for beers at Churchill's Royal Canadian Legion Hall. When I arrived, I found a long table of men in Carhartt overalls and quilted flannel shirts decompressing after work. The national news was on in the corner, and the men hollered and shushed one another when a report about Martha Stewart's visit came on the television. Everyone paid attention, but hardly anyone said anything when it was over. They mostly just groaned or huffed air through their noses.

I'd been spending a lot of time hanging around the entrepreneurial second tier of Churchill's tourism industry, getting to know many of the same weather-beaten men who'd helped pioneer the industry twenty and thirty years ago. I liked them. And, I'll admit it, I enjoyed it in a very high-schoolish way whenever I walked into Gypsy's Bakery and got waved over to sit with them at the table in front of the pastry case—the table always understood to be held just for them, even during the mightiest mealtime rush of classic tourists. But none of them believed climate change was endangering the town's polar bears, no matter how clear-cut the science seemed to me, and this had a way of stunting conversation. No one was worried, and they brushed the prospect aside so casually, with such genial sarcasm, that I sometimes felt embarrassed about bringing it up. As one man put it to me: "Yeah.

The bears are all fucking dropping through the ice, because it's melting and drowning them. Right."

Nevertheless, many of them could rattle off changes in the climate that they'd noticed with their own eyes. The summer was clearly longer. The vegetation was bigger. Moose were more common. So were robins. ("It used to be a really big deal to see a robin up here," one guy said.) In fact, the main economic hope for the town these days is that a longer season of open water on Hudson Bay will bring more ships into Churchill's seaport and that, eventually, the town will become Canada's outpost for trade with Russia, with ships loading up with grain from the prairies and navigating a new, ice-free Northwest Passage.

Still, nearly all the locals I met were convinced this warming was just part of a natural cycle. And the presumption was, if the change in the climate were natural, as opposed to man-made, the polar bears were somehow inherently equipped to withstand it. Maybe the bears' numbers would shrink, but they'd muddle through until the climate corrected itself and there was plenty of ice again.

Everyone came back to an image of the polar bear as an invincible eating machine. Many of these men spend a good part of their year on the land, hunting and trapping; they know the bears don't technically "fast" through the summer months, as biologists tend to put it. "The polar bear eats," Mark Ingebrigtson, a former mayor of Churchill, told me. "He gets the odd seal, the odd whale." People told me about bears they'd seen tearing through berry bushes. They'd seen them pillage goose nests and inhale the eggs. One man described watching a polar bear lie in wait while a caribou herd blew past, then pounce on the last animal in line and devour it. Another claimed to have watched a bear leap from the bank of the Churchill River onto the back of a beluga and drag the whale to shore like a Labrador fetching a floppy, inflatable pool toy. A male bear can easily weigh twelve hundred pounds. It can run twenty or twenty-five miles an hour and be at full speed after only a few strides. They are opportunists, I was told, and can eat virtually anything they want. Even if the ice recedes, they are not going to "stand around starving" onshore.

Recently, some of these same adaptation theories have been argued out in scientific journals.* But the consensus remains that the polar bear, as a species, simply can't last by haphazardly grabbing calories on land. It takes a huge amount of energy to keep a polar bear operating. (According to one study, if a polar bear chased a goose for more than twelve seconds it would have burned more calories than it gained by eating the goose.) It's not a coincidence that the animals have evolved to eat ringed seals—hapless globs of fat, which they hunt by lying perfectly still next to a seal's breathing hole and waiting for it to surface. Everything else is just snacks. Individual, innovative bears or pockets of bears may manage to outlast others for a while, but the ice

---

* For example, a snow goose biologist at the American Museum of Natural History, Robert Rockwell, who has worked in Churchill for more than forty years, made a case in one paper that the town's polar bears could actually survive on land by subsisting on goose eggs. Polar bear biologists rebutted his work, and thus began an esoteric back-and-forth involving population dynamic modeling and hypothetical calorie crunching. It was boring, but each new paper made headlines in the press, which has taken to covering polar bear science as though it were a boxing match between environmentalists and climate skeptics. Rockwell's study was seized on as a damning counterfactual by skeptics, while environmentalists branded him a climate denier. (He's not.)

This was only one of several dust-ups in the press around the time I visited Churchill. A polar bear scientist was investigated (and cleared) by the government for exaggerating reports of drowned polar bears in Alaska. A photo of a polar bear clinging to a small fragment of ice was exposed as an artful crop job. People seemed to believe that disproving any particular claim about polar bears and their vulnerability to climate change invalidates the reality of climate change altogether—even if, as in Rockwell's scenario, the ice is still melting and the bears are merely skirting by on a goose-egg technicality. Rockwell, for his part, wound up feeling that most of the people who accused him of being a climate skeptic were "idiots" and that the scientists who lashed out to challenge his research, including a couple of big name polar bear biologists affiliated with Polar Bears International, have become so desperate to convince an unresponsive public about climate change that they're behaving irresponsibly. They exaggerate the facts and are intolerant of any science that complicates their tidy storyline, Rockwell told me. They mean well but, they're talking like "marketing guys" now, not scientists.

is slipping away too fast for the species to re-engineer its metabolism and evolve. When ice receded after the last ice age, a population of polar bears around southern Scandinavia died out in almost exactly the same way.

But the argument in Churchill wasn't really scientific anyway. It was, I realized, a more philosophical fight about what the animal is capable of, the character of the beast. People in Churchill live in the bear's world and see polar bears do clever and creative things. It's impossible not to anthropomorphize the animals, to see them as having a humanlike capacity for problem solving that makes them, like us, nearly invincible, and to assume they'll adapt, even if we never thought to expect such ingenuity from a less impressive critter like the copper-striped blue-tailed skink, a lizard in Hawaii, when the ecology it was a part of changed. (The skink was last seen on Kauai in the 1960s; in 2012, the government officially declared it extinct.) In other words, biologists recognize the polar bear as just one cog in a Darwinian machine—one that will drop out when the structure holding it up deteriorates. But people in town see it as a menacing and capable agent of its own fate. It was obvious that they expected the same resourcefulness and perseverance out of their polar bears that they themselves showed after the military left.

On top of that, so many people in Churchill that I met were also not inclined to believe the climate change story because they resented the messenger. They saw Polar Bears International as carpetbaggers from down south, an elite NGO that set up shop every fall and flew in a pageant of scientists and overachieving American high-schoolers, and capitalized on Churchill's polar bears without much meaningful interaction with Churchill itself. At the Legion Hall that night, everyone was quick to point out the obvious synergy between PBI and the tour company Frontiers North: the more PBI muscled the media into talking about climate change, the faster tourists would pay Frontiers

North for the chance to see the animals before they disappeared. Money was presumed to be changing hands somehow, and virtually everyone at the table was convinced that Robert Buchanan is secretly a part owner of Frontiers North—that he had engineered the whole masterful scheme.

The real victim was the bear, they told me. "All I know is that I'm tired of people feeling guilty about coming up to see polar bears. It's absolutely unfair to the bears," said Kelsey Eliasson, a contemplative younger guy with gnarled facial hair. He'd come to Churchill in 1999 as an idealistic environmentalist—he owned a composting toilet and a van that ran on vegetable oil. "I believed," he told me. "I was a believer. I was an annoying eco-freako." But he was turned off by the way he saw activists like Polar Bears International "Disney-fying" the bear. They played up its vulnerability, twisting it into a cuddly sob story, and it made Kelsey wonder if climate scientists spun the truth in similar ways. "Now I don't even know whether to believe in climate change anymore," he said.

A guy a few seats down, who'd been silent, reached around to pat Kelsey on the back. "We always believed somewhere down the road that you'd come around," he said. His name was Dennis Compayre. People called him Dennis the Bear Man. He had longish, sandy hair, and resembled a thick-set Dennis Hopper. He was imposing, but enunciated all of his words perfectly with the breathy, faintly patrician voice of a Haverford comp-lit professor.

Dennis had driven Tundra Buggies from the very beginning, in 1982, until his friend Len Smith sold the business to Frontiers North. He loved polar bears and, after parting ways with the new ownership, still wanted to be able to spend bear season on the tundra with them. So, in 2000, he dug his old rig, Buggy One, out of a boneyard, rebuilt the machine, and soon talked Frontiers North into going into business with him. He installed a webcam on Buggy One's roof, drove it into the viewing area, and camped in it all fall, beaming grainy footage of bears and keeping an online diary for a few hundred paid subscribers.

Early one morning during Dennis's second autumn, Dancer came knocking—thundering on the hull of Buggy One like a friendly drunk who'd lost his keys. Dancer was a huge male polar bear who'd hung around the early photography tours in the eighties. That was the lard-tossing era, before feeding bears was illegal, and by feeding Dancer, Dennis had taught the bear to walk backward on his hind legs—to dance. "It's a terrible thing, I know," he told me. "Like a circus act." Twenty years later, Dancer seemed to remember Dennis and Buggy One. He immediately started doing the dance. And from then on, and every bear season for the next several years while Dennis ran his "bear cam," Dancer would trail right behind Buggy One like a loyal hound. The media loved it, and the man and bear gained an international following. Then, after the 2005 season, Frontiers North leased Buggy One to Polar Bears International. Dennis was out. Only so many vehicles are allowed on the tundra, and, given the urgency of the climate crisis, Dennis—this feral Lebowski of a man, throwing the occasional sausage to the dancing polar bear behind him—did not seem to be leveraging that access in a productive or meaningful way. His audience was small. He was off-message. As Robert Buchanan put it to me, "His webcam didn't have much of a consumer franchise."

Polar Bears International overhauled Buggy One again, rebuilding its interior so that, in a matter of minutes, it can be transformed into a mobile television studio. Now during bear season PBI and its partners fly in a different panel of polar bear and climate specialists every week and broadcast live chats from the tundra to schoolkids and college classrooms, or to jam-packed special events at zoos. PBI has branded these brief talk-show-style programs Tundra Connections.

I felt bad for Dennis. I'd heard he could be a disagreeable man, but, still, he'd lost his pet project. And he felt something else was being lost, too. "People come up here now with a lump in their throat because they think this bear is doomed," he told me. "Not for the joy of being with a bear, and seeing a bear in the wild. That's secondary now." The animal was being diminished, disparaged. He couldn't conceive of the need to grease a wild animal into a sympathetic environmental "franchise," just as, thirty years earlier, people in Churchill couldn't understand why National Geographic felt obligated to frame the bear as a cage-rattling monster. It was as though, the closer you were to the actual bears, the harder they were to square with any one story—even, in the case of climate change, with the truth.

EXTINCTION IS NOT AN EASY IDEA to get your head around. Centuries ago, as the first colonists hacked away at the wilderness of the eastern seaboard, they sometimes noticed that the deer or wolves they were encroaching on and killing disappeared rapidly from the immediate area. But the possibility that a species could be annihilated totally, everywhere, was literally inconceivable: it occurred to almost no one.

Partly, this had to do with the way people imagined America. In such an infinite-seeming space, brimming with wild things, surely there'd always be somewhere else for these displaced animals to go. But it also had to do with how people imagined nature. All species were believed to be part of a "Great Chain of Being"—a sturdy hierarchy into which God had ordered and fixed all living things, from bugs and slugs all the way up to angels. The idea that any part of God's perfect chain could be destroyed was both illogical and sacrilegious. "Such is the economy of nature," Thomas Jefferson wrote, "that no instance can be produced of her having permitted any one race of her animals to become extinct; of her having formed any link in her great work so weak as to be broken." I was finding a vestige of this idea around Churchill—the attitude that, as long as the change in the climate was natural, it couldn't threaten the bears. Nature, Jefferson argued, is not in the business of driving its own animals extinct.

Jefferson was writing in the early 1780s, in a book called *Notes on the State of Virginia*. He was defending his decision to include the mammoth in a preceding list of contemporary American animals, even though no one had encountered a mammoth alive. Mammoth fossils—one-pound teeth and femurs larger than my daughter—had started to be unearthed nearly a hundred years earlier, first in upstate New York and later from an area in Kentucky dubbed Big Bone Lick. Confronted with mammoth fossils, Jefferson, like others, saw no reason to believe that herds of these giant animals weren't still grazing across the "unexplored and undisturbed" interior of the continent. "Our entire ignorance of the immense country to the West and North-West, and of its contents, does not authorise us to say what it does not contain," he argued. Years later, he'd advise Lewis and Clark to keep their eyes peeled for these monsters. Americans had begun calling the mammoth the American Incognitum, Latin for "unknown."

Jefferson makes a big deal of the Incognitum in *Notes on the State of Virginia*. He spends several pages waxing about the animal's hugeness and fussily discrediting claims that the bones being dug up are actually from less impressive, more common animals. He sounds defensive—petty, even. And I came to understand it's because the argument Jefferson was trying to settle wasn't just about Incognitums. He was marshaling the mammoth forward as a symbol, a stirring icon that—not unlike the polar bear today—might have the power to change public opinion about a critical issue of his time.

Jefferson was trying to debunk the Theory of American Degeneracy, which had been worked up several years earlier by the revered Enlightenment writer and natural historian, Count Georges-Louis Leclerc Buffon. Count Buffon argued that the animal life of the New

World was smaller, weaker, and less spectacular than that of the Old World. Not only were there fewer species in America—less of a diversity of life—but individual animals there were smaller than their counterparts in Europe: they were "degenerate" versions. Buffon himself had never been to America, it turns out, so he plucked measurements to support his claims from the accounts of travelers, who weren't always reliable. (At one point in his book *Natural History,* Buffon repeats a claim that the region around Hudson Bay was populated by pygmies, winged serpents, and savages with backward-bending knees.)

The degeneracy theory turned on a question of climate. Buffon believed that species could attain their ideal forms only in a warm climate. He imagined America as a humid stew of swamps and uncultivated wildness; he believed its landmass had only recently emerged from under the sea and was still drying out. The continent, he wrote, is "crisscrossed by old trees laden with parasitic plants, lichens, [and] fungi, the impure fruits of corruption." Eventually, Buffon concluded that even lines of domesticated animals brought to America from Europe gradually shrank and degenerated. The dogs in America were "absolutely dumb." The sheep didn't taste as delicious.

Others picked up on Buffon's theory. Soon the fact that America's birds or squirrels were smaller than Europe's became a stand-in for the irredeemable smallness of everything else in America. The critiques became snider and snobbier, and strayed further from Buffon's original quantitative claims. ("A stupid imbecility is the fundamental disposition of all Americans," one writer noted of the Indians.) Soon champions of degeneracy were claiming that there was something so vexingly degenerate about America that Europeans who immigrated there would have inferior kids—that their bloodlines would degenerate just like the livestock's. They noted America's conspicuous lack of great men. Europe had produced Socrates, Copernicus, and so on. But, as one writer pointed out, "Through the whole extent of America there had never appeared a philosopher, an artist, a man of learning whose name had found a place in the history of science or whose talents have been of any use to others."

As ridiculous as it might sound, the degeneracy idea gained enough traction that it threatened to undercut America's standing in the world. Confidence in the new American nation was already shaky. Europeans doubted whether the crowd of rowdy idealists we now call the Founding Fathers could hold the whole project together. Buffon's degeneracy theory now gave an empirical basis for those doubts. It made America look like an inherently bad bet, and might dissuade already skeptical nations from loaning the United States money. In short, this wasn't just a fight about their foxes being bigger than our foxes. Something had to be done. Thomas Jefferson stepped up.

It's easy to see why the degeneracy myth rankled a man like Jefferson. Jefferson was almost unbearably left-brained, a stickler for precise and provable facts. He recorded the daily weather for forty-four years and measured his own walking speed (four miles and 264 yards per hour). He also appears to have been thin-skinned and a little vindictive. At the Continental Congress, as delegates slashed sections from his original draft of the Declaration of Independence, he sulked in his chair, looking wounded; Benjamin Franklin tried to loosen up his friend by muttering little jokes under his breath. (Jefferson felt so hurt by how the congress had "mangled" his writing that he later issued a kind of director's cut of the Declaration to his friends, with all his original text restored.) That Buffon and his followers were passing off their bunkum as science would have infuriated him as an empiricist as well as a patriot.

In an excellent book about the episode, biologist Lee Alan Dugatkin explains that disproving Buffon became "one of Jefferson's great obsessions"—and one that would escalate, to the point of absurdity, through some of the most otherwise productive years of his life. Jefferson started sensibly enough, by attacking the credibility of Buffon's data and gathering data of his own. (It became a national project. Ben Franklin helped out, as did James Madison, who sent a letter to Jefferson reporting the dimensions of a particular weasel. Madison's letter was detailed, down to the "distance between the anus and the vulva.") But before long, Jefferson's counteroffensive became less straitlaced. Buffon had written that America had no panthers, only smaller and less impressive cougars. Jefferson wanted to show the count that America *did* have panthers, and that American panthers were very formidable. So, en route to France, where he was going to serve as ambassador, Jefferson impulsively bought a panther skin and took it with him, to deliver to Buffon and prove his point. The Frenchman wrote Jefferson to thank him for the gift. But he referred to it as a "cougar" skin, not a "panther" skin. Buffon did promise to correct the mistake in a future edition of his book—not the theory of degeneracy as a whole, just the panther bit. Nevertheless, Jefferson could be found bragging triumphantly about this nitpicky panther comeuppance in a letter to a friend forty years later.

After he arrived in Paris, it took Jefferson nearly a year to get a face-to-face meeting with his adversary. When Jefferson finally had an opportunity to confront Buffon, at a dinner the count hosted, he was cut off; Buffon handed him a new manuscript, saying, "When Mr. Jefferson shall have read this, he will be perfectly satisfied that I am right." Later, over dinner, Buffon said something flippant about American deer, rousing Jefferson to insist that American deer are awesome and have horns that are two feet long. Buffon also let slip that he didn't believe in moose. He presumed this allegedly new American species to be merely a form of reindeer that some degenerate American naturalist had incorrectly given a new name. Jefferson spoke up, telling Buffon, as he later recounted it, "The rein deer could walk under the belly of our moose." Buffon laughed this off. So Thomas Jefferson started issuing a flurry of letters home, pleading for someone in America, anyone, to kill and stuff the largest moose he could find and ship it to him in France.

The moose became a fixation. Dugatkin writes that for the better part of a year, "in the midst of correspondences with James Monroe, George Washington, John Adams, and Benjamin Franklin over urgent matters of state, Jefferson found the time to repeatedly write his colleagues—particularly those who liked to hunt—all but begging them to send him a moose." But the moose that finally arrived in France a year later seems to have been more than a little pathetic. A military captain in Vermont, subcontracted by a former governor of New Hampshire, had finally succeeding in killing a seven-foot-tall moose for Jefferson. But the moose fell deep in the wilderness, twenty miles from the nearest road. A team of men had to clear a path through heavy snow and haul the animal out. The carcass decayed, the skin wasted, and the meat putrefied during the fourteen-day-long trek. It was, the governor wrote to Jefferson, a "very troublesome affair" and he was "much mortified" by the expense. (He included receipts so Jefferson could reimburse him.) The moose's antlers had to be tossed, so the governor sent along sets of deer, elk, and caribou antlers instead, for Jefferson to affix to the head or mix and match as he saw fit.

Jefferson sent the moose to Buffon. He did his best to talk it up. He swore to the count that, although this moose had gone mostly bald in transit and its remaining fur was now shedding, it definitely had a full and glorious coat when killed. He also apologized for the elk horns, which were on the small side. "I have certainly seen of them which would have weighed five or six times as much," he insisted.

Ultimately, it's easy to imagine Thomas Jefferson as an early American George Costanza, a seething nebbish quick to take umbrage but never quite able to respond convincingly. The theory of degeneracy would go away only gradually as, in response, Americans turned that chauvinism on its head and told a more compelling story about themselves. They'd begin celebrating the raw wildness of their country: America as a land full of big and beautiful things, and Americans as a people tied to nature's rhythms—farming and hunting rather than sitting in European parlors. "Nature," one writer claimed, "was establishing a system of freedom in America" that Old Europe "could neither comprehend or discern."

It is impossible to know what impact Jefferson's moose had on Buffon's thinking. Not long after receiving it, the Count died.

IN *NOTES ON THE STATE OF VIRGINIA,* Jefferson compiled the data he'd gathered—measurements and weights of animals in Europe and America—and presented a table of side-by-side comparisons: our 410-pound bear versus their 153.7-pound bear; our 12-pound otter versus their 8.9-pound otter. Having argued, at length, that the American Incognitum shouldn't be presumed extinct, he put it all by itself, at the top, to give the table a little wow factor.

Notes on the State of Virginia helped elevate the mammoth into an icon of patriotic pride. The historian Paul Semonin calls the animal "a symbol of overwhelming power in a psychologically insecure society." Its story hinted at the wonders the new continent might contain. The mammoth was gossiped about at the Continental Congress, and several of the Founding Fathers collected its fossilized teeth. As president, Jefferson made his own paleontological study of the beast, laying out a set of bones on the floor of a room in the White House like a child's electric train set.

In 1802, during Jefferson's first term, the first complete mammoth skeleton was mounted in a blockbuster exhibit by the Philadelphia museum owner Charles Willson Peale. "A kind of mammoth fever swept the nation," as Semonin puts it. A Philadelphia baker made a

"Mammoth Bread." Two butchers sent Jefferson a "Mammoth Veal," the hindquarter of an especially elephantine, 436-pound calf. (Jefferson declined to eat the veal, which had been shipped long-distance, but praised it as an example of "enlarging the animal volume.") In Washington, a "Mammoth Eater" shoveled forty-two eggs down his throat in ten minutes. When Peale's son took the mammoth skeleton on a tour of Europe, the museum hosted a "Mammoth Feast" to send him off. A dozen guests ate at a banquet table under the Incognitum's rib cage while a pianist played "Yankee Doodle."

Meanwhile, scientific evidence that the Incognitum was extinct mounted. And in the end, there was no getting around the logical assumption that, if these fossilized animals did still roam the earth, humankind would have stumbled into them by now. It was through the example of the mammoth that the concept of extinction gained credibility, undoing people's belief in the Great Chain of Being.

Americans, though, were slow to accept that fact—until it, too, was given a more palatable spin. Writers framed the disappearance of the mammoth as a divine blessing on America. Clearly, Americans couldn't claim dominion over the continent if it were teeming with angry mammoths. And so God had wiped out "this terrible disturber." The idea of extinction had undermined religious belief in the Great Chain of Being. But now it was reinterpreted to confirm the central myth of a new religion: Americanism. God had purged the mammoth so that the young nation could spread out and absorb the empty continent ahead of it.

This was an extinction America could get behind—evidence of our singular and overpowering status in the world. The likelihood of the polar bear's extinction tells the same story. But we don't feel good about it this time.

3.

BILLY POSSUMS

n 2007, a fourth-grader at Sobrante Park Elementary in Oakland, California, wrote a letter to the head of the United States Department of the Interior. "My name is Juan Piedra," he began. "Every morning when I wake up I tell myself how much danger the polar bears are in right now, and how sad I am right now. Imagine if everything was upside down. Please help the polar bears. I am really heart broken. They are feeling badly."

I found Juan's letter in an archive of public comments submitted to the government after it announced that it was considering putting the polar bear on the endangered species list. Normally, these decision-making processes are quiet, complicated ordeals, hashed out by bureaucrats and lawyers; calling for comments from the public is a pro forma part of the procedure. But with the polar bear, half a million supportive letters, postcards, and petitions from Americans poured into the Department of the Interior—then the most ever in the Endangered Species Act's history. Many were handwritten pleas from children to "save the polar bear," and some offered solutions to climate change, like using ethanol instead of fossil fuels. (A kid named Fritz wrote: "I feel bad about the polar bears. I like polar bears. Everyone can use corn juice for cars. From, Fritz.") Lots of kids just drew pictures: polar bears wearing life preservers, or stuck on little ice islands, or—in one case—a polar bear drowning and being eaten simultaneously by a shark and a lobster.

The polar bear was, by this time, a pop culture preoccupation—"a shining white symbol of the green movement," as one television news reporter put it. The mania was sparked in early 2005, when environmental groups, led by the nonprofit Center for Biological Diversity, had first petitioned the Department of the Interior to consider endangered status for the bear, setting the legal procedure in motion. Petitions to list species as endangered are filed all the time. In fact, the Center for Biological Diversity and another group, WildEarth Guardians, would soon be filing them at a combined rate of about three hundred per year—pressuring the federal government to protect all manner of imperiled sturgeon and bats that most environmental groups don't bother to lobby for. But the Center for Biological Diversity hoped the polar bear might stir up special interest. It could be a landmark case: the first species protected explicitly because of the threat of climate change.

The morning after the center filed its petition, MSNBC splashed a picture of a polar bear across its home page. Then CNN did. A long polar bear blitz began. Over the next several years, public attention to climate change intensified, spurred on by events that often had nothing to do directly with polar bears. The summer of 2005 saw less ice cover in the Arctic than any other summer since satellite monitoring began three decades earlier. In 2006, Al Gore released *An Inconvenient Truth.* Then, in 2007, a high-profile panel of scientists convened by the United Nations released its final, sobering report about the "un-equivocal" certainty of climate change and its projected effects. ("This is real, this is real, this is real," one of the lead authors said, explaining the findings to the press.) But because the problem of climate change was invisible, the media found that polar bears were an easy, adorable means to illustrate these stories—more eye-catching than a smokestack spewing carbon or a glacier crumbling. *Time* magazine ran a photo of a bear on its cover with the headline "Be Worried. Be Very Worried." Annie Leibovitz photographed Knut, a celebrity polar bear cub at the Berlin Zoo, with Leonardo DiCaprio for the cover of *Vanity Fair.*

The species had become a spokes-species, and no matter what context polar bears appeared in, they symbolized the same thing. It had gotten to the point that, by the end of 2007, New Line Cinema, the makers of the fantasy film *The Golden Compass,* which featured a computer-generated, armored white bear as one of its characters, worked with the World Wildlife Fund to produce public service announcements about climate change using clips from the film. They also donated several hundred thousand dollars to the conservation group. It was as though New Line were paying a karmic licensing fee for the use of a white bear, even though *The Golden Compass* never actually mentioned polar bears and took place in another universe. "This is a very organic partnership for us," a New Line marketing executive insisted.

This convulsion of polar-bear love may look like an empty craze. But all that visibility, and all those children's letters, had real, political consequences. Even though the fervor for polar bears wasn't engineered by the Center for Biological Diversity, they were counting on it as part of a plan they'd laid out in advance. They were using the bear as a trap in a much bigger and longer-running legalistic war of words.

In the early 2000s, environmental attorneys were struggling to force the Bush administration to regulate greenhouse gases. The Bush administration, meanwhile, kept refusing even to acknowledge definitively that those emissions were causing climate change. In 2003, the legal tack that had seemed the most promising—that carbon should be controlled as another kind of pollution under the Clean Air Act—got stalled in the courts. Looking for new angles of attack, two attorneys at the Center for Biological Diversity, Kassie Siegel and Brendan Cummings, turned to the Endangered Species Act.

The Endangered Species Act lays out a program for the conservation of imperiled plant and animal species. It makes it possible to devote money and government workers to their recovery and to set aside and protect land they live on. It bans killing, harassing, or shipping those species across state lines or overseas, and forces government agencies to make sure that their activities—everything from building a new fence to testing bombs—don't endanger them further.

The government's decisions about which animals deserve to be on the endangered species list must be based solely on the "best available science"—whatever studies have been conducted that speak to the severity of the threat of their extinction. It wasn't clear how far listing the polar bear could go toward actually saving the species; some significant steps could be taken under the Endangered Species Act, but it seemed unlikely that any administration would upend America's entire carbon-based economy to fulfill its technical obligations to the polar bear under the law. But petitioning the Bush administration to rule on whether the polar bear *qualified* for protection would at least confront the government, and the public, with the climate science it had so far managed to duck—a first step to any eventual progress. It was a way to put the government on the spot; the polar bear and the entire Endangered Species Act were being played like pawns in a higher-stakes chess match. In fact, Siegel and Cummings had come up with the strategy several years earlier and had already auditioned other species for the role of climate change victim.

The science of climate change was well understood at the time, but there still weren't many published studies showing how specific species would be affected—and, for the environmentalists' strategy to work, the "best available science" the Bush administration was going to be cornered with needed to be ironclad. Siegel and Cummings were left scraping the bottom of the taxonomic barrel. They considered using the Glacier Bay wolf spider, a spider in Alaska. But there was uncertainty as to whether the wolf spider was a distinct species, and whether it therefore qualified for protection. Also, the Glacier Bay wolf spider sounded icky, a public relations nonstarter, unlikely to focus the American public, and not just the courts, on climate change as the case picked up steam.

In 2001, Siegel and Cummings petitioned the government on behalf of the one truly solid case they could find: the Kittlitz's murrelet, a little-known speckled Alaskan seabird that frequently nests near shrinking glacial ice sheets, and whose population—estimated in the low tens of thousands of birds—may have declined more than 85 percent since 1991. The outcome of the Kittlitz's murrelet petition was discouraging. The Bush administration didn't exactly deny the bird endangered status, but it didn't protect it, either. The Kittlitz's was deemed "warranted but precluded." It was shoved through a curious loophole in the law onto a backlog known as "the candidate list."

The candidate list has a complicated history. The modern Endangered Species Act was passed in 1973, by an overwhelming majority of senators and congressmen. It announced itself as a counterforce to the "consequences of economic growth and development untempered by adequate concern." And, although this may sound radical—the United States government pledging to temper the country's growth— it was treated as feel-good, softball stuff at the time. A law to save animals was a relief from Watergate and Vietnam. Its passage was hardly noted—the *Washington Post, New York Times,* and *Los Angeles Times* each devoted exactly one sentence to it—and President Nixon signed it into law in the doldrums between Christmas and New Year's.

One historian writes that most in Congress believed the Endangered Species Act was "a largely symbolic effort" to protect only the kinds of species environmentalists call "charismatic megafauna"— grizzlies, whales, bald eagles, and other large, beautiful species that people tend to feel an easy connection with. But the act had been quietly beefed up by idealistic staffers, and was much further-reaching and more powerful than most congressmen took the time to understand. After its passage, there was instantly a lot of buyer's remorse. (In a famous example, a small fish called the snail darter quickly complicated a dam-building project in Tennessee.) Meanwhile, protection was being sought for obscure birds and skinny little snakes that none of these legislators had ever heard of. Within two years, some twenty-three thousand species had been proposed for endangered status. The Smithsonian pulled together a list of 3,187 plants it considered worthy of protection. The paperwork alone was staggering. The agencies responsible for ruling on those petitions, primarily the Fish and Wildlife Service, soon found ways of brushing them off their desks.

Congress reformed the law several times to make the listing process more functional and fair. It required the government to rule on petitions according to strict timelines, but also set up the candidate list so that, if a particular species needed protection in a hurry, Fish and Wildlife would have the flexibility to deal with that crisis. Designating a species "warranted but precluded" would put it in a temporary holding pen, pausing the clock on its petition deadline, while the agency made progress on its more pressing work. But the workload involved in staving off extinction and managing all of America's endangered species only grows, and in our new era of conservation reliance, it appears to be open-ended. Especially since the 1990s, the government has used the warranted but precluded category as an indefinite dumping ground.

In 2005, the Center for Biological Diversity found that many candidate species had been waiting around on the list for an average of seventeen years, some of them holdovers from that original list of imperiled plants drawn up by the Smithsonian. Recently, the center and other groups sued the federal government to spring those species from their bureaucratic purgatory and give them full protection. The government settled and will now be slowly reassessing each one's case. But the settlement didn't close the warranted-but-precluded loophole. It's likely only a matter of time before the candidate list starts filling up again.

Around the time I visited Churchill, there were nearly three hundred species on the candidate list. I found a copy of the list and noticed that virtually all the species on it have one thing in common: I've never heard of them. There's the Neosho mucket mussel and the Slabside pearlymussel, the band-rumped storm petrel, spotless crake, relict leopard frog, smalleye shiner, and least chub. The Roy Prairie pocket gopher is one of nine pocket gophers on the list. There are several bats, five kinds of salamanders, nine snails, and four shrimps. There's the Sonoyta mud turtle and Miami blue butterfly; the Clifton Cave beetle, the Coleman Cave beetle, the Fowler's Cave beetle, the Indian Grave Point Cave beetle, the Icebox Cave beetle, the Inquirer Cave beetle, the Louisville Cave beetle, the Nobletts Cave beetle, and the Tatum Cave beetle. Also, Stephan's Riffle beetle. And there are plants, like Hirst's panic grass and Short's bladderpod. A Hawaiian plant called the Alani spent fifteen years on the candidate list before it was finally bumped up to endangered status in 1994. Unfortunately, the plant appeared to have gone extinct two years earlier.

At least twenty-four species seem to have gone extinct while waiting around on the candidate list, and I've never heard of them, either. They include a fish called the shortnose cisco and lots and lots of species of mussels, including one called the lined pocketbook. In 1982, something called the Valdina Farms salamander, which lived in a single cave in Texas, was deemed warranted but precluded. Five years later, a water agency diverted a river and flooded the cave, wiping it out.

This was the real value of the polar bear, then: its magnetism. It could inspire enough public gushing to make it politically impossible for the Bush administration to dump it quietly onto the candidate list and bury the issue of climate change yet again. The public-relations strategy was also a legal strategy. As the Center for Biological Diversity's Brendan Cummings put it at the time, "No politician wants to tell their kids, tell their constituency, 'Yes, I voted to kill the polar bear.'" The Endangered Species Act may say that we, as a nation, are devoted to preventing the extinction of any more species. But we also know that we can't realistically save everything. And no one cried for the lined pocketbook.

WHY ARE WE DRAWN to certain wild animals and not others? Can the cultural carrying capacity of a species—its charisma, essentially—be predicted or deconstructed?

That's the mystery that the Center for Biological Diversity was trying to game in its listing petitions, and that conservation groups have long puzzled over, working to move the public to a particular ecological cause through the story of just the right, sympathetic victim—the bald eagle, which brought attention to DDT in the 1970s; or the spotted owl, which took on the logging industry in the nineties. Why exactly, according to one survey, are 73 percent of Americans willing to block construction of a power plant and pay more for their electricity in order to save mountain lions, but only 48 percent willing to do so to protect a plant called Furbish's lousewort—especially since, frankly, few of us are likely ever to see a mountain lion or a Furbish's lousewort whether the power plant is built or not.*

Part of the answer seems to be that we are attracted to animals that resemble us physically, a principle called "phylogenetic relatedness." Monkeys are more likable than otters; and otters—with their recognizable facial structures, little mustaches, and shrunken hands—are more likable than lizards. We may be especially sympathetic to phylogenetically related animals because we assume that a creature that

---

* Over the last twenty years, a new field of academic study has coalesced around similar questions, examining our attitudes toward animals and the sociological, psychological, and imaginative forces that influence them. The field is so new that its own researchers don't always agree on a name for it—often it's called Human-Animal Studies—and its findings are wide-ranging. My favorites include: The more television a person in upstate New York watches, the more fearful he or she is of being attacked by a black bear. Americans are more likely to assume that a given tiger is female than male. If the mammals depicted on beer bottle labels reflected the actual mammalian biodiversity of Earth, there would be far more rodents and bats on beer bottles, and far fewer mountain goats, bighorn sheep, and wooly mammoths. Only 13.7 percent of white women in Southern California won't enter the ocean out of fear of jellyfish. American television commercials tend to depict solitary wild animals, whereas Chinese ads show herds, flocks, and gaggles. The average time a person spends in front of an animal enclosure at an American zoo is 99.31 seconds.

In a study in which a fake snake, a fake turtle, and a Styrofoam cup were placed on the side of a road, motorists hit the snake and turtle more often than the cup, and the snake more often than the turtle; nearly 3 percent of motorists who hit the fake animals appeared to hit them on purpose. Another study, looking at people's reactions to being attacked by pumas, found that the "lowest likelihood of escaping injury occurred when individuals remained stationary." Women are more likely than men to get "a magical feeling" when seeing dolphins in the surf. Sixty-eight percent of "mothers with high feelings of entitlement and self-esteem" identified with a dancing cat in a commercial for Purina.

Americans consider lobsters more important than pigeons, but also more stupid. Turkeys are seen as slightly more dangerous than sea otters, and people believe dolphins to be smarter and more lovable than human beings. Pandas are twice as lovable as ladybugs.

looks vaguely like us will have similarly high capacities for thought, pain, and feeling. (In one study, researchers told interviewees that a small mob had just cornered and kicked an animal "like a football" until it was bloodied, unconscious, or dead. The more similar that animal was to humans, the stiffer the fine or the more jail time interviewees recommended for the abusers.)

We are also evolutionarily programmed to empathize with species that resemble human babies—with large, forward-facing eyes; floppy limbs; circular faces; and a roly-poly shape. This helps explain, for example, why polar bear cubs wind up on so many cutesy wall calendars, and why cartoon fish, like Pixar's Nemo, are never drawn realistically, with eyes on either side of their heads. The Yale social ecologist Stephen Kellert has summed it up this way: "People generally prefer large attractive animals with an erect bearing, animals that walk, run, or fly rather than crawl, slither, or live underground. A good candidate for the average human nightmare might be a creature that is small, ugly, predatory, likely to inflict injury or property damage, lacking in intelligence or feeling, and a denizen of dark, damp places, inclined to crawl and slither about." In other words, we like the polar bear, not the Glacier Bay wolf spider.

Still, physicality explains only so much, and what it does explain can feel obvious. There is a purely cultural dimension to the way we think about wild animals; their meanings can shift and float in and out of fashion over time. As the softening of the polar bear's image suggests, the stories we tell about animals depend on the times and places in which we tell them. This was proved more than a century ago, during an inadvertent nationwide popularity contest of bear versus opossum.

It began in November 1902, when President Theodore Roosevelt took a train to Mississippi, to escape the White House for four days of roughing it and black bear hunting outside the town of Smedes. On the second morning of the hunt, the dogs caught the scent of a bear and chased it into the swampy thickets outside of camp. After a chase, Roosevelt turned back for lunch. But his hunting guide—a yarn-spinning ex-slave named Holt Collier, well-known in the Delta for having killed three thousand bears—eventually managed to corner the animal near a watering hole late that afternoon. The bear snatched one of the hounds by the neck and mashed its spine, killing it. After it injured a second dog, Collier leapt off his horse and cracked the bear on the head so hard that he bent back the butt of his rifle. Then he roped the animal to a tree and tooted away on his bugle, calling in the president for the honor of the kill.

The bear was a 235-pound female—semiconscious, injured, mangy-looking by some accounts, and, Collier judged, shrunken to about half its normal weight by Mississippi's drought. When Roosevelt saw the pitiful animal lashed to the tree, he refused to fire at it, or to have anyone else shoot it, either; he felt it went against his code as a sportsman. Instead, he asked a hunting companion to put the bear out of its misery with a knife. But that detail of the story would quickly get lost. A few days later, a political cartoonist in Washington, Clifford Berryman, memorialized the moment when Roosevelt declined to fire his weapon as an almost saintly scene. He called the cartoon "Drawing the Line in Mississippi." Roosevelt was shown with his rifle down and his hand outstretched to spare the bear, while the animal sat on its hind legs like a baying puppy, with frightened wide eyes and two ears pricked up on the top of its head. It looked as helpless as an infant, as if it needed to be reassured or swept into someone's arms. It wouldn't have registered as familiar at the time, but, looking at the cartoon now, you recognize the animal right away: it's a teddy bear.

Essentially, the bear from the cartoon was turned into a plush toy and named after the president. There are competing legends about who made the first teddy bears: it was either Rose Michtom, the wife of a Brooklyn toy-shop owner, or a German seamstress named Margarete Steiff, whose family owned the felt manufacturing company Steiff, still the world's most prominent teddy bear producer. We do know that Steiff had been selling a line of stuffed animal toys, including a bear, for several years before Roosevelt's hunt. But Steiff's original bear was a much more realistic animal, less cuddly and infantile, with the humping, brutish back of a wild one. Also, the bear was chained through its nose to a peg.

Bears, after all, were considered monsters. For so long, the animal had been a shorthand for the unruliness and danger that Americans were encountering on the western frontier. Bears rarely turned up in toy catalogs and books, one historian notes, and "when they did they looked mean and were apparently designed to upset young children." Two years before Roosevelt's trip, *Ladies' Home Journal* published a kids' adventure story about a fourteen-year-old named Balser, described as "the happiest boy in Indiana" because he owned a rifle, "ten pounds of powder, and lead enough to kill every living creature within a radius of five miles." In the story, Balser winds up killing a bear, but gets bitten in the process. So, in the story's feel-good conclusion, the boy and his father track down the bear's mate and shoot her, too, in revenge.

For bears—real bears, out on the land, with pulses and appetites—turn-of-the-century America was a painful and inhospitable place. All kinds of large carnivores were being systematically exterminated, from east to west, to keep from complicating the lives of humans. Wolves, cougars, and coyotes especially were demonized as Americans' competitors: "brutal murderers" that killed and ate "harmless, beautiful animals"—namely, the livestock that people were raising to eat themselves. In 1906, an arm of the federal government, the Bureau of Biological Survey, began killing tens of thousands of wolves and coyotes every year, with traps and poisoned meat. The government also offered bounties, roping ordinary citizens into the work. One bureau biologist would justify the war on wolves by insisting, "Large predatory mammals, destructive of livestock and game, no longer have a place in our advancing civilization."

This is to say, the teddy bear was born in the middle of a great spasm of extermination that would go on for decades. (Even the Audubon Society began eradicating predatory birds, like hawks and eagles, from their bird sanctuaries.) It was a natural escalation of the mind-set formed a century earlier, in Thomas Jefferson's time, when Americans told themselves that the gruesome Incognitum had been driven extinct to wipe the continent clean for their use. Now the country was finishing off all these smaller, less imposing Incognitums—buffing out the land's last scratches of wildness so that all we could see in its surface was our own reflection.

The teddy bear was only one sign that some people, deep down, had started to feel conflicted about all that killing. America still hated and feared the bear. But all of a sudden, America also wanted to give the bear a hug.

THIS AFFECTION was already starting to percolate when Roosevelt went to Mississippi. Two years earlier, in 1900, the bestselling author Ernest Thompson Seton published *The Biography of a Grizzly*, a book that tenderized the reputation of the bear in the same way the teddy bear would. The story begins with a mother grizzly and her cubs "living the quiet life that all bears prefer." But when a rancher opens fire, only one cub survives—a morose little guy named Wahb who must find his way in a shrinking wilderness riddled with steel traps and tainted by the "horrible odor" of man. Yes, Seton argued, grizzlies were once ferocious. But the barbarity of men with rifles and traps had put them in their place. Now was the time to show the bear some mercy: "The giant has become inoffensive now," he later wrote. "He is shy, indeed, and seeks only to be let mind his own business."

By the time Seton wrote *The Biography of a Grizzly,* he was a controversial figure at the vanguard of a new literary genre called realistic wild animal stories. These stories claimed to be credible natural histories of wildlife. But they dramatized the lives of animals as though they were the anthropomorphic heroes of fiction. (Jack London's *White Fang* may be the realistic wild animal story that's best remembered today.) Seton insisted that his stories were steeped in a nuanced and accurate knowledge of animal behavior, gained from his years in the field. And yet he endowed his animals with a cleverness and morality that sometimes border on the ridiculous. He wrote, for example, of a mother fox that feeds her trapped offspring poisoned meat so that the pup won't have to suffer the indignity of being chained up. Then she nobly commits suicide herself.

Seton was not the most unrealistic realistic wild animal story author. Some almost completely sanitized nature of its violence or trauma. (The writer William Long described a scene of wolves ripping apart a deer as being "peaceable as a breakfast table.") Still, Seton was one of the most successful authors, and he became a target for the backlash against the genre by other naturalists. One critic derided realistic wild animal stories as the "yellow journalism of the woods." Theodore Roosevelt was one of the authors' most vicious enemies, dubbing them "nature fakers," which is the name by which they're remembered today. The fear was that these writers were misleading readers about the way nature worked. Children would be especially vulnerable to their lies.

The country was urbanizing. By 1910, a majority of Americans would live in cities. Instead of spending time in nature, children relied on secondhand descriptions of wildlife now, and naturalists worried that, without much firsthand experience of animals, kids might accept even these sappy bedtime stories as fact. Teachers around the country were starting to use another of Seton's books, *Wild Animals I Have Known,* as a textbook. "All of this would be highly amusing," one zoo director wrote, "if it were not so pitifully serious to the children of the public schools."

But some of the nature fakers' motivations were more poignant than their critics understood. Seton especially was responding to America's war on predators. His most dignified, sympathetic protagonists were usually the same animals that were being exterminated in the West, like grizzlies and wolves. He was trying to create public empathy for these species—to save them. Like Robert Buchanan with his polar bears more than a hundred years later, Seton knew that regurgitating dry, scientific descriptions wasn't enough to generate a true emotional response. Seton's aim instead, he wrote, was to capture the "personality" of an individual animal "and his view of life." "Since, then, the animals are creatures with wants and feelings differing only in degree from our own, they surely have their rights."

The nature fakers may be mostly forgotten, but this sentimental compassion lives on in nearly every children's book about animals I've read to my daughter—books that, like everything adults give to little children, are echoes of our own beliefs. And it was evident, too, in so many of the letters about polar bears that schoolchildren wrote to the Department of the Interior in 2007. "I really think it is not fair to the polar bears," wrote the fourth-grader in Oakland, Juan Piedra. "Also, they could drown and die off and what if they were you?"

Nature can seem this pure and honorable only once we're no longer afraid of it. We seem to be forever oscillating between demonizing and eradicating certain animals, and then, having beaten those creatures back, empathizing with them as underdogs and wanting to show them compassion. We exert our power, but are then unsettled by how powerful we are.

Large predators—those able to rip us apart—have understandably commanded a huge share of humans' psychic attention for as long as there have been humans. (Some of the earliest cave paintings are of bears and lions.) But as we've insulated ourselves from nature, and diffused the danger of those animals, we've started to give them new meanings. That basin of anxious, imaginative energy can get rechanneled into a deep aesthetic appreciation. In the bear especially, Yale's Stephen Kellert argues, we see a creature a lot like us: it can walk upright, snores when it sleeps, and is roughly our size and shape. But it's also omnivorous, agile, clever, self-possessed—all the admirable dimensions of ourselves that have been "diminished in modern culture." For many of us today, who spend our days slumped over spreadsheets or quarreling with our banks over hidden fees, bears look like the composed and competent survivors we wish we still were.

No single piece of research demonstrates this cycle of fear and reverence more clearly than a study, led by the geographer Jennifer Wolch, that examined how cougars were written about in the *Los Angeles Times* between 1985 and 1995. In the early 1970s, the cougar population in California had been ground down to as low as twenty-four hundred animals. But by 1990, a ban on hunting had allowed the species to come back; the cougar had become an icon of conservation in Southern California. It was described in the newspaper as "majestic" and "innocent," an embodiment of nature's grace, and a "symbol of our dwindling wilderness heritage." But soon cougars started encroaching into the populated areas around Los Angeles. There were two fatal attacks. More people still died in America because of bee stings and black widow spider bites, Wolch writes, but "as reports of cougar-human interaction rose and public fears were fanned by episodic attacks, the images of cougars as charismatic and proud wild animals at home in nature were replaced by terms conjuring danger, death, and criminal intent." It was as if a switch had flipped. Before 1990, the predominant image in the newspaper was of an "elusive and fascinating wild creature." After 1990, cougars were "efficient four-legged killers" and baby-snatchers, "roaming like phantoms" in the nearby hills.

The same shift has been happening with wolves lately, especially since Republican legislators maneuvered via a last-minute budget amendment to take away the gray wolf's federal protection in several states in 2011. (Conservationists defended the wolf as part of America's natural majesty; Montana's governor, meanwhile, told his constituents to forget the Endangered Species Act altogether and take matters into their own hands: "If there is a dang wolf in your corral attacking your pregnant cow, shoot that wolf. And if its pals are in the corral, shoot them, too," he told Reuters.) And a decade after Wolch's cougar study, similar research looked at newspaper editorials about a proposed black bear hunt in New Jersey and found almost exactly the same scenario: bears being cast both as "menacing threats" and as "God's creatures" who would gladly "live in peace" if people just left them alone.

When Roosevelt refused to shoot that black bear in Mississippi in 1902, the species' larger cousin, the grizzly, was being brutally eradicated around the country. And as it disappeared from the land, it found new prestige in our imaginations. Soon a novel by James Oliver Curwood, called *The Grizzly King: A Romance of the Wild,* would turn on a scene that is almost exactly the opposite of what happened on the president's bear hunt. A grizzly named Thor stalks the hunter who has previously shot and wounded him. The bear creeps in behind the hunter, trapping him between a rock wall and a cliff, with nowhere to run—and unarmed. Thor towers over the man angrily, but then pauses, stunned by how "shrinking, harmless and terrified" the creature that had hurt him looked now. And so the animal slowly turns and disappears in the direction from which he'd come, leaving the hunter standing there—letting him live.

The bear was now the merciful one, with a code of honor he refused to break. The hunter was the senseless killer. As Seton once wrote, "No animal will give up its whole life seeking revenge; that kind of mind is found in man alone. The brute creation seeks for peace." The bear was the bigger man.

IT DIDN'T TAKE LONG after Roosevelt's bear hunt in 1902 for the teddy bear to become a full-blown craze. By the end of the decade, Steiff was producing close to a million teddy bears a year. Sets of teddy-bear clothes were sold separately, and *Ladies' Home Journal* published patterns for making your own. Your teddy bear could wear pajamas or dress up like a sailor or a fireman. There were even special blankets and caps to keep the toys toasty in winter. That is, despite all its fur, the bear needed a winter coat. In the natural history of the teddy bear, this seems to be the point at which the teddy bear splintered into its own discrete species, when it completely broke away in our imaginations from its relative in the forest.

But the toy confused adults. Their children were trading in dainty baby dolls for beasts—it was troubling. "From all quarters of the globe," wrote the *Washington Post,* "comes the demand for Teddy bears, with poor Miss Dolly gazing woefully out of her wide open eyes powerless to prevent the slipping away of her power." The *New York Times* published a poem: "The Passing of the Doll." The teddy bear seemed like a novelty—a fad—and everyone assumed it would be forgotten once Roosevelt left office. Mass-manufactured toys themselves were still fairly new, and so, as the inauguration of Roosevelt's successor, William Howard Taft, approached in 1909, the toy industry was hungry to ramp up production of America's next cuddly plaything— whatever it might be.

That January, President-elect Taft was the guest of honor at a banquet in Atlanta. The big news, for days in advance, was the menu. The Atlanta Chamber of Commerce was going to serve Taft possum and taters, a Southern specialty that one writer of the time described as "the Christmas goose of the epicurean negro." An opossum, roasted on a bed of sweet potatoes, was typically presented whole—head on, pale tail hanging off it like a meaty noodle—with a smaller potato crammed between the animal's fifty tiny teeth. The one brought to Taft's table weighed eighteen pounds.

After the meal, the orchestra started to play, and the guests suddenly broke into song while Taft, presumably caught off guard, was presented with a gift. It was a small stuffed opossum toy, beady-eyed and bald-eared. This brand-new creation was intended by a group of local boosters as the William Howard Taft presidency's answer to the teddy bear. They called it the Billy Possum.

A company, the Georgia Billy Possum Co., was already being formed in Atlanta for large-scale manufacturing of these stuffed animals. According to one account, deals for Billy Possums were being brokered with toy distributors across the country within twenty-four hours of the banquet. (It seems that the company initially experimented with stuffing actual opossum skins, but wound up with something too fleshy-looking and repulsive—like a pale, limp rat.) The *Los Angeles Times* covered the unveiling of the new toy at the Chamber of Commerce banquet and announced, "The Teddy Bear has been relegated to a seat in the rear, and for four years, possibly eight, the children of the United States will play with 'Billy Possums.'"

A fit of opossum fever began. There were soon Billy Possum postcards, Billy Possum pins, and Billy Possum pitchers for cream at coffee time. There was even a new ragtime tune: "Possum: The Latest Craze." Real opossums weren't that common in cities. So a toy shop in Brooklyn ran an in-store promotion with a live, captive opossum, so that children could familiarize themselves with the animal that was primed to "rival the Teddy Bear in popularity." ("Do not let it be said," the store's advertisement read, "that any man, woman or child in Brooklyn has not seen the cute little animal whose name is mentioned more perhaps in all parts of the world to-day than any other.") At Taft's inaugural parade, the Georgia delegation was given Billy Possums to wear clipped to their lapels. There were smaller Billy Possums–on-a-stick to wave like flags.

But, despite all this marketing, the life of the Billy Possum turned out to be demoralizingly brief. The toy was a flop, peaking and petering out within months of its introduction that January and almost entirely forgotten by the end of the year. That is, Billy Possum never even made it to Christmastime, a special sort of failure for a toy.

In retrospect, the failure of the Billy Possum can probably be explained two ways. The first is straightforward: opossums are ugly. But the Billy Possum's backstory was all wrong, too, particularly compared with the teddy bear's.

Through most of human's evolutionary history, what has made the bear magnificent in our eyes is the animal's independence from us— its parallel life as a menace and competitor. But by the time Roosevelt was hunting bears in Mississippi, with the country exterminating its predators from coast to coast, that stature was being crushed. That one black bear, tied to a tree outside Smedes, symbolized the predicament of all bears. The animals now lived or died according to our wants and whims. It said something ominous about the future of bears, but it also raised disquieting questions about who we'd become, if the survival of such a creature was now up to us. The legend of Roosevelt and the bear resonated as an allegory of the confusion that

America was only beginning to face. The bear was a helpless victim roped to a tree. The president of the United States decided to show it some mercy.

Taft, on the other hand, ate his opossum for supper. He ate a lot of it, in fact—so much that, after his first several helpings, a doctor seated nearby actually passed him a note, suggesting it might be a good idea if he slowed down. "Well I like possum," Taft told reporters the next day. "I ate very heartily of it last night, and it did not disturb in the slightest my digestion or my sleep."

Today a small selection of stuffed opossums has found its way back onto the market. Judging from the reviews I found on Amazon, the toys seem to be mostly bought as gag gifts for people who have had creepy run-ins with actual opossums. One woman explains that the Fiesta Toys ten-inch plush opossum is so realistic-looking that her daughter screamed when she first took it out of the box. "We all love it now," the woman goes on, "but opossums are not lovable in real life."

A CENTURY AFTER ROOSEVELT drew the line in Mississippi, the trap that the Center for Biological Diversity was setting for the Bush administration hinged on one question: Was the polar bear a teddy bear or a Billy Possum? How lovable was it? Now that it had been proposed for endangered species status, would the animal whip up enough public sympathy to steer clear of the candidate list and force the administration's hand, or could it be quietly shunted aside like the Kittlitz's murrelet? In the end, the answer was more complicated than anyone imagined.

In 2008, the Bush administration did place the polar bear on the endangered species list. It classified the bear as "threatened," a desig-nation that gives the government more flexibility and doesn't guarantee the same level of protection for the species that a fully "endangered" one receives. This allowed the administration to write what's called a 4d rule for the polar bear, an amendment that adjusts how the law will apply to a particular species. The polar bear's 4d rule was exceptionally dramatic. It asserted that regulating greenhouse gases was outside the bounds of the Endangered Species Act; in this one case, the Fish and Wildlife Service was exempt from addressing the primary threat to an imperiled species. In a press conference, Secretary of the Interior Dirk Kempthorne explained that he wasn't about to let a law about animals be "abused to make global warming policies." The government, finding no way to wiggle out of the corner that the Center for Biological Diversity had backed it into, had looked the environmentalists right in the eye, kicked a ragged hole in the wall, and crawled through it.

An almost incomprehensible carnival of lawsuits kicked off. The Center for Biological Diversity and its partners ginned up several. The first, brought against the government, tried to get rid of the 4d rule by demanding that the bear be listed as endangered and not just threatened (4d rules can be applied only to threatened species). They presented rather embarrassing internal government documents showing that the decision had been politically manipulated, not solely based on the best available science. This, in turn, forced the government—it was the Obama administration by now—to defend the bear's threatened classification. Surprisingly, at no point in the history of the Endangered Species Act had anyone had to parse the legal difference between "endangered" and "threatened," and so the government now produced a richly perplexing document that tried to do just that, drawing ephemeral distinctions between phrases like "on the brink of extinction" and "the step just prior to the brink of extinction" that allowed it to define "threatened" in a way that applied perfectly to the polar bear's situation. Of course, this semantic hair-splitting then had to be rebutted by the Center for Biological Diversity, which offered its own semantic hair-splitting.

By the time I visited Churchill, the whole legal fight had, to my mind, devolved into an existential debate about the nature of time. (If the government defined "endangered" as likely to go extinct, then "threatened" must mean likely to be likely to go extinct. But what does *that* mean? And so on.) As the litigation vanished deeper into this procedural rabbit warren, the media lost interest. It got hard even to remember the Center for Biological Diversity's original goals: to get America thinking seriously about climate change; to get the Endangered Species Act and the entire national project of conservation that it enables to start addressing, or just acknowledging, climate change as the game-changing, environmental challenge of our time; to begin to imagine how it will undermine or downright shatter the work of conservationists who, having fought to keep imperiled species swaddled safely inside their native habitats, will now watch the habitats themselves change, or fall out from under the animals entirely, like the sea ice under the bears.

The polar bear, really, was just a prop to underscore the problem of climate change—a problem that, if left unaddressed, begs the question of whether addressing anything else is worthwhile. But now everyone had been yanked into a frothing, bottomless argument about the prop itself. Six years after she'd filed the original petition to list the bear, the Center for Biological Diversity's Kassie Siegel was in a federal court in Washington arguing the definitions of "endangered" and "threatened" again when, finally, the judge asked her: "What does all that mean in the real world?"

4.

THE CONNECTION

One afternoon, I rented a truck and drove outside town to see a dog breeder named Brian Ladoon. It was a bleak day, even for Churchill. Clouds lumped in the sky like smoke, and wind charged off the crumpling gray slate of the bay.

Ladoon was born in Churchill and has lived here most of his life. The previous week, he'd finished third out of three candidates in Churchill's mayoral election. (His platform was to shut down the town every winter, charter a big jet, and fly everyone somewhere tropical. He got thirty-five votes.) He's played a big role in reviving a rare breed of dog, called the Canadian Eskimo dog, and keeps his stock of about 140 animals on a sloping tract of coastline, far off the area's one actual road, behind the old military radar domes. People call the area Mile 5. The dogs are chained to stakes down near the water, where a dirt trail empties into a wide bulb of rocky land. I could hear them baying and howling when I pulled up. Offshore, the rusted wreck of a ship called the *MV Ithaca,* which ran aground in 1960, tilted out of Hudson Bay. It was an austere scene, and as I took it in through the windshield of my truck, a school bus full of tourists suddenly pulled away and a preposterously big polar bear came into view behind it. The bear was walking alongside a blue hatchback, dwarfing it.

Every fall, a gang of male bears, said to be the largest in the population, hang out at Mile 5, attracted by the dog food. They roam the rocky spit while the dogs cluster together and bark and shriek to challenge them. Ladoon charges tourists who come to see the bears—sometimes $40 per hour per person, I heard; sometimes just a bottle of rum. During bear season, he's here every day, patrolling the dirt trails in his black pickup, chain-smoking Marlboros, and doing his best to keep the animals from getting too close to the cars of his customers. In town, an acquaintance of Ladoon's had told me, "He thinks he can talk to the bears. He thinks they understand him."

Ladoon was expecting me. He got out to undo the chain he keeps slung between two posts as a gate. Then, motioning for me to stay put in my vehicle, he turned around, unzipped his pants, and took a piss. He is fifty-seven and was dressed all in black, with dark, narrow eyes, a white goatee, and long white-silver hair that was kept matted against his head by a black leather headband. Eventually, he cleared off his passenger seat and I got in.

The place is essentially a tumbledown, drive-through safari, and the disorderliness of it was only heightened by this imperturbable guy in the weird headband claiming to have everything under control. "I make sure the bears don't molest the people, and the people don't get themselves grabbed," he told me as we resumed his rounds. Soon a bear rose up from the roadside and walked toward his side of the truck. "That's a twelve-hundred-pounder right off the hoof, eh?" Ladoon said casually. He slapped his horn twice, but the bear kept coming. When it got within a few feet, Ladoon leaned out his window, revved his engine, and shouted at the animal. His stoner drawl exploded into a deep, low growl. He said, "No, you asshole!"

At that, the bear dropped its head. Instantly, all menace drained out of the animal. I watched it lope away, lie on its stomach, cross its paws into a cushion, and slump its head in them. The bear kept eyeing us, but it looked chastened, like a dog who'd been bad. It was amazing. "He knows," Ladoon said softly. "He knows." Often, he told me, all he has to do is pump the action on his shotgun and most bears will back up at the sound of it—he's got them "trained." Ladoon grinned at his weapon in between our two seats and the rounds of cracker shells and rubber bullets. "They know there could be anything coming after that," he said. "You know what I mean? 'Here comes the salad, boys! First course!'"

It was an open secret in town that Ladoon was also keeping the bears in check by feeding them, which is illegal. He adamantly denied it. But even friends of Ladoon's, like Paul Ratson and Dennis Compayre, discussed this with me freely. (Ladoon also explained his bear-feeding regimen to *Canadian Geographic* magazine in 1997.) Biologists I met in Churchill tended to regard him as a low-life bear-baiter who uses his dogs as a front to keep collecting payments at his gate; they worry that he's teaching the bears to associate humans and dogs with food, which will lead to more encounters and conflict. But many locals just see Ladoon as an eccentric, if unflattering, fact of small-town life. Churchill's mayor, Mike Spence, suggested I bring my daughter, Isla, to Mile 5 when she and my wife got to town. "It's like a Sunday drive through the park, so to speak," Spence said. "She'll be amazed at how big the bears are."

As we drove, Ladoon explained that wildlife photographers and camera crews have been coming to photograph and film bears at his dog yard since the eighties. He bragged that some of the world's most recognizable polar bear pictures, including several magazine and book covers, originated here, and began to explain why the particular access to polar bears he provides is so invaluable.

As photographers discovered Churchill in the eighties and nineties, they also discovered stock agencies and magazines with a large appetite for their polar bear pictures. But eventually, with the advent of digital photography, it no longer took skill to capture a white bear in a white landscape at the right exposure—anyone could do it. By now, one photographer told me, "Polar bears have been photographed to death." So many photographers shot bears in Churchill that they'd nearly obliterated the demand for those pictures, just as hunters can shoot so many animals they obliterate the supply. More important, photo editors found that they could afford to be picky, and that so many of the pictures pouring out of the town looked identical and somehow wrong. They all peered down on polar bears from the high deck of a Tundra Buggy, minimizing the animal. And their backgrounds were laced with Tundra Buggy tire tracks and dirt roads, spoiling the image of polar bears as lonely rogues in a wide and desolate wilderness.

At Ladoon's, though, the bears turned up just as reliably as in the viewing area where the buggies go, and could be photographed more intimately, at eye level, or even looking up into their harrowing faces. And the property looked pristine and varied. Ladoon is an artist himself—he used to devote a lot of time to painting. From his truck, he started pointing out to me the different backdrops that he offers to photographers: the shoreline, the frozen ponds, the bear trails that wind through the willows. He was the curator of all these real-life landscape paintings for the polar bears to wander in and out of. "There's so many theaters," he said, "so many dynamics that can happen in each theater." What had looked to me a minute ago like bleak and formless nature now resembled a Hollywood back lot. Here photographers could capture polar bears exactly as the public expected to see them.

The more professional photographers I met in Churchill, the more

I realized that a good wildlife photograph or film, or at least a market-able one, does just this: shows us an image of nature that's already lodged in our heads.* However, our imaginative sense of an animal is so powerful that it can also change what we see in pictures.

The German photographer Norbert Rosing first met Brian Ladoon in 1988. He would spend virtually every bear season in Churchill for the next twenty years, often photographing the bears at Ladoon's dog yard. (Rosing told me that he's never seen Ladoon feed the bears in-tentionally but that the polar bears clearly swipe their share of the dogs' food.) Late one afternoon in 1991, Rosing watched a bear slink hesitantly toward one of Ladoon's dogs. Its posture went soft. It lofted its right paw in the air, toward the top of the dog's head, like an old man patting a child. Gradually, the dog got comfortable and ap-proached the bear. Soon the two animals were hugging—actually hugging—with the dog straining on its chain to nuzzle its neck against the bear's, and the bear enclosing the dog with its fluffy forearm. La-doon had told Rosing about this bear, which periodically turned up to

---

* It takes extreme amounts of time, money, patience, and luck to catch that sort of iconic material in the wild, and, understandably, some professionals cut corners. Chris Palmer, a veteran wildlife filmmaker who recently authored an exposé of the industry, explains how animals from game farms are routinely used as stand-ins for wild ones, or jelly beans are hid-den inside deer carcasses so that trained bears will tear them apart. This kind of trickery has been going on forever. (In 1958, Disney wanted to show lemmings scrambling, en masse, off a cliff in the Arctic for its film *White Wilderness*. So the Disney crew paid Inuit kids to round up lemmings, forced the lemmings to run on a treadmill covered with snow, then picked up the lemmings with their hands and chucked them into the water. In reality, scientists later determined, lemmings don't even run off cliffs. Americans still think they do in large part because *White Wilderness* popularized the idea.) But, Palmer argues, the explosion of nature programming on television, with dedicated twenty-four-hour networks like Animal Planet and Nat Geo Wild, has only made things worse. It's created a demand for more, and more sensational, footage but shriveled filmmakers' budgets and deadlines. Wildlife filmmakers, Palmer told me, are good people and often staunch conservationists, but the pressure on them is agonizing, and the ethical lines are blurry. "It's not that you're evil or malignant or mali-cious," he said. "You're just trying to get the damn shot so you can go home and have dinner with your family. So you put the monkey and the boa constrictor in the same enclosure."

"play" with the dogs. But Rosing hadn't believed him. The animals carried on until, finally, the bear was sprawled on its back in the snow, peering up, gazing into the dog's eyes.

In 1994, Rosing sold a series of photographs to *National Geographic*, documenting this entire play session. Immediately, he was besieged by angry faxes and phone calls. The public image of the polar bear was still what it had been a decade earlier, when National Geographic broadcast *Polar Bear Alert*: a fierce killer that terrified mothers in the middle of the night and assaulted cameramen in cages. People assumed the dog had been chained up as bait for the white monster— clearly, the bear wasn't playing, but springing a sinister trap; it must have gored the dog right after Rosing's last shot. No one wanted to see the photos, Rosing told me. "People just couldn't believe it." After a while, he put the pictures away.

Thirteen years later, the pictures found their way onto the Web site of a public radio show in Minnesota. It was the summer of 2007 now—polar bear fever, brought on by the endangered species list petition, was peaking. The bear had been transformed in people's minds. It was adorable now, defenseless—less like a marauder and more like a teddy bear, an animal that *would* be inclined to play. "Now people feel they can touch and pet bears," Rosing told me, "because they're just so nice, so cute, so curious." And because the polar bear looked different, the pictures looked different, too. The same photos that had reviled people in 1994 now touched them. They rapidly racked up three million views on the radio show's Web site, then spurted around the Internet, where they've cheerfully blossomed in all kinds of contexts since.

Recently, a friend forwarded me a chain e-mail he'd received from a woman he described as "literally a friend of a friend of a friend's grandmother" in the Midwest. Inside the e-mail were Rosing's photos of the bear and dog, cuddling in a corner of Ladoon's dog yard. "It's hard to believe this polar bear only needed to hug someone!" the e-mail read. "May you always have love to share, health to spare, and friends that care."

It was a Thursday—Ladoon had forgotten—and that meant that his hired hands, two young guys named Caleb and Jeremy, had run into town to fetch the week's dog food: thirty-three hundred pounds of frozen chicken necks and by-products. There were slabs of the stuff, each the size of a small tabletop, stacked in the back of the boys' pickup. It needed to be transferred to Ladoon's vehicle. They also had a dog to chain back up with the others.

It was a high-wire act. At one point, as Jeremy stepped out of the truck cradling the dog to his chest, a polar bear began galloping down the road, attracted to the smell of the food, probably, or of the dog. Or of Jeremy. Ladoon had to jerk our vehicle into reverse to intercept it. As he swung my side of the truck in front of the animal, I saw gelatinous ropes of slobber swinging from its mouth. The bear stopped short, groping for another angle. Ladoon only nodded disapprovingly and said, "He's off the bear-ometer."

When it was all done, we returned to the gate and found a blue SUV idling between Ladoon's chain posts. The driver had slipped through without paying. The same animal that had made a go at Jeremy now stood a few feet away, perched with all four paws contracted under the fat jumble of its body, like a circus animal posing on a barrel. A white-haired woman in the passenger seat was taking pictures of it. I watched as she began to lean through the open window just slightly, extending her lens, then her face, through that last intangible boundary between her space and the bear's.

"Who the hell are these guys?" Ladoon said. He was about to scold them when the polar bear straightened up and took a single, vaulting step toward the woman, instantly cutting the distance between them by half. The woman reacted late—very late. I watched her hands flub around under the window for whatever button or crank would shut it.

Now the bear skirted around to face the SUV head-on. It stood on its two back legs and raised its front paws. Then it leaned forward and fell, its paws thwacking into the hood. "That's a rental vehicle," Ladoon said. His voice was perfectly measured, as though he were thinking about only what a headache this would be for the woman who owned Churchill's rental-car business. And yet Ladoon was simultaneously revving his engine, cranking his pickup into a shuddering W-shaped turn, and hammering his horn.

The road was so narrow that he had no room to maneuver and scare the animal away. So he gunned straight ahead and kept driving, slapping his hand against the outside of his door to lure the bear and clear the area. Turning quickly, I saw it and another polar bear clomping after us. Then I heard a thud and felt our truck bobble on its suspension.

One of the bears had hurled itself onto the back of our truck. It was going for the blocks of frozen chicken in the backseat. "He almost got it." Jeremy laughed. "He got one tooth on it, but slipped!"

Ladoon fumed. He clearly wanted to do some hollering. But by the time he drove back to the gate to scold the people, the SUV was gone. "These bears aren't cute," he later explained. "Look how big these fuckers are! Everybody wants to get close to the bears. Well, there's a time to get close to the bears and there's a time to—you know—maybe stay *far away* from the bears."

Eventually, I found out who the older woman in the SUV was: Margie Carroll, an ebullient, retired schoolteacher from Georgia who'd come to Churchill to sell copies of her self-published children's book *Portia Polar Bear's Birthday Wish*. I met Carroll later, when I was invited to dinner at one of Polar Bear International's rental houses in town; Carroll was a friend of the organization and was staying there. After dinner, she scurried upstairs and returned with another of her books, *A Busy Spring for Grandella the Gray Fox,* and read most of the book to our end of the table out loud, in her melodious Southern accent, rambling into extemporaneous asides when they occurred to her. On one page, for example, the father fox brings his children something to eat. The text reads, "Daddy is such a good hunter," and all the characters peek hungrily down at the ground near the father's feet. Carroll pointed out how she'd stopped short of showing the mangled prey; the characters only stare at blank space. "I wanted to show the family unit nurturing the children," she explained, "but I didn't want to have a bloody bunny saying, 'Help me!'" She understood that nature is violent, but felt that violence worked at cross-purposes here. Kids need to see that animals' lives are enriched by the same family values as their own, she said, and that wildlife therefore deserves our compassion. She was an heir to Ernest Thompson Seton, in short. "I wanted to show children to value nature. Don't just go kill something! It's part of a family," Carroll said. Then she flipped to the next page.

Portia Polar Bear's Birthday Wish is an equally syrupy and meandering tale about a polar bear cub who feels very insecure about her crooked feet. Her birthday wish is "to be normal." Carroll told me that Portia's experience is supposed to show kids that they shouldn't obsess about their tiniest imperfections; it's okay to be unique. It's a parable, trying to universalize the woes of one particular animal, she explained. In that way, I realized, it wasn't so different from the story that Polar Bears International is telling to adults. I bought a copy for Isla and asked Carroll to sign it.

Reading it at bedtime one night, I found myself thinking back on the incident at Mile 5, how gripping it felt in retrospect: how the actual polar bear rose up to threaten the author of *Portias Polar Bear's Birthday Wish,* punishing the hood of her car. It was as though the bewildering distance between something imaginary and something real had finally collapsed, if only for a second.

My wife and daughter arrived in Churchill early on a Wednesday morning. The one-room airport terminal was thick with bear tourists who'd streamed off the same flight. Among them was a stylish Canadian late-night host named George Stroumboulopoulos. Strombo, as he's known, was coming to Churchill with two Canadian rock stars to film a polar bear special with Polar Bears International after Martha Stewart's crew was done. My wife, Wandee, had noticed Strombo posing for photographs with all the stewardesses and airport gate attendants in Winnipeg and assumed he was merely an uncommonly enthusiastic tourist, documenting every leg of his trip.

Isla was a little over two years old. It had been ten days since I last saw her. I'd left before dawn on the morning after Halloween, having taken her around the neighborhood the night before. I wore the Winnie-the-Pooh costume she'd decided at the last minute not to wear, stretching its hood around my face so that the yellow bear-body swung in front of my chest like a fleecy beard. Isla went as an eggplant with wings. Seeing me at the airport now, she tucked her head into her shoulder and cemented her face into an unimpressed glare— my punishment, it seemed, for being gone. But then something occurred to her and she flung out one leg, showing me the pair of blue long johns under her pajamas. (She'd never worn long underwear before.) I yanked up the leg of my jeans and made a big show of revealing that I, too, was wearing long underwear, and her face broke wide open into a grin. With that, we seemed to have worked through any hard feelings.

My trip to Churchill became a working vacation now. I was free to be the same sort of sentimental polar bear tourist that so many of the folks I'd been meeting in town felt ambivalent about. The truth is, I'd arrived to see polar bears with the same jumbled baggage as did other classic tourists—emotions I couldn't quite sort out and, frankly, wasn't comfortable delving into. The fact that something as large and autonomous-seeming as a polar bear might stop existing, and the even more tremendous fact of climate change, made me viscerally uneasy whenever I allowed myself to truly think about them. So I usually didn't. And it was this very detachment that troubled me most—how easy it was to watch the future of the planet, Isla's future, spool ahead like a negligible fiction. I couldn't do much to stop the disappearance of polar bears. But I figured that the least I could do was force myself to pay it some very serious and deliberate attention—to put myself, and my daughter, near some sign of the upheaval under way.

I worried that Isla wouldn't do well trapped on a Tundra Buggy for an all-day tour, so I'd finagled the three of us spots on a half-day supply run out to Frontiers North's Tundra Buggy Lodge, a chain of stationary buggies in the center of the viewing area that the company has remade into an inn, with bunks and a galley. (A five-night package at the lodge can cost close to nine thousand dollars; the chance to see polar bears wander by the window over one's morning coffee, before the other buggies arrive, commands a premium.) But at some point on our school bus ride from the hotel to the buggy launch site, those plans suddenly fell through. Before I knew what was happening, we were bounced onto a buggy with the staff of Polar Bears International and the winners of its annual Project Polar Bear competition. Climbing aboard, we found a hive of amped-up young people and a very drowsy-looking Robert Buchanan crumpling himself into one of the backseats. "These are our carbon footprint–reducing whiz kids," he said.

PBI has run the Project Polar Bear competition since 2008. It sets teams of teenagers around the world to the task of identifying and implementing ways to reduce greenhouse gas emissions in their communities. The team that corks up the most carbon wins a trip to Churchill. That year, there were two winning teams. The "Canuck Nanooks" were a family affair: the four polite and baby-faced Vickery sisters, who ranged in age from fourteen to twenty, and hailed from a rural hamlet outside Winnipeg. The other team, from Louisville, Kentucky, called itself "There for Tomorrow."

These were Polar Bears International's youngest, most heartfelt "Arctic Ambassadors." There was a patch that said as much on the sleeves of their matching blue PBI parkas. "Ambassador" was a word that Robert took very seriously. I'd noticed during the past week that he has a catalog of little stump speeches he draws on when talking to journalists. One of the key ones describes his mission to convert "tourists" into "ambassadors." Over breakfast one morning, Robert had laid the speech on me, then stopped to unpack the meaning of the word.

Being among wild polar bears on the tundra has the potential to truly reshape people, he told me. "I call it 'the Connection,'" he said. "The ultimate Connection is when someone is able to look in the bear's eyes. That bear will reach into your heart and your soul, and you are changed forever." The Connection isn't tree-hugging fluff, Robert went on; it's Marketing 101. Making eye contact with a bear "screams," as they say in the ad game. It screams the way the lavish, omni-colored box of Froot Loops screams to a kid in the grocery aisle. It grabs you and touches you on a level beyond intellect, demanding your compassion. An "ambassador," Robert told me, is someone who makes the Connection and then goes home committed to helping that polar bear and its compatriots in the wild. A tourist is just looking to be entertained. Tourists go home and only start planning their next vacation. "But if you're coming all the way up here just to see polar bears so you can check them off your list," Robert said, "if you're coming for your own self-interest, do me a favor: don't fucking come. You can see it on the Discovery Channel." The carbon emissions generated by flying to Churchill are astronomic, he said. You have to offset them with action.

Now, as our buggy jolted over slushy potholes, the teenagers seemed to whirr in their seats with an anticipatory high. They snapped photos of each other flashing peace signs, and then double peace signs, and invented and practiced handshakes in which peace signs turned into the wriggling antennae of a snail. Some of the younger Vickerys sang. There was now at least a thin dusting of snow on the ground, and what's known as grease ice had finally started clumping on Hudson Bay, undulating with the tide beneath it, like the skin on a cup of cream of mushroom soup.

We found our first bear only a few minutes outside the launch site. Wandee held up Isla so she could see, and I wedged myself in behind them. The bear lay with its neck stretched forward and its eyes closed. One of its flanks was tinted red by the rising sun. The sight of the animal abruptly silenced everyone on board. As people shifted to that side of the buggy and drew down the windows, all you could hear was the quiet whizzing and clicking of digital cameras. This hush was familiar. I'd been on several buggy rides, and it happened at every first sighting, either out of reverence for the animal, or just because everyone was so instantaneously stupefied by the spectacle of an actual, wild polar bear that it didn't occur to them to talk. But this time, the quiet of the congregation was broken by my daughter, shouting shrilly into the wind. "Wake up, polar bear!" Isla said.

ISLA MADE A POINT of talking to several polar bears that day. She was still feeling her way into language, and her crimped way of expressing herself reminded me of how, in nature documentaries, a just-born giraffe will skitter on its matchstick legs before finding its footing. She had lots of questions: whether polar bears liked jelly; why she couldn't go out there with them. At one point, we watched a lone animal stand up and retreat. The bear wasn't in good shape. Its ribs were visible faintly under its fur, like a name materializing in a gravestone etching.

Isla said, "Whoa."

"'Whoa' is right," said Sam Leist, an irrepressibly likable kid on the Louisville team.

Then Isla said, "Where do bears poo-poo?" and I watched Sam pull a befuddled, kindly half-smile, searching for any possible point of entry, before turning back to the window and raising his binoculars again. Frequently, my daughter just shouted, "That bear's too big!" with a mix of disapproval and disbelief.

I'd been prepping Isla for the trip with polar bear footage on YouTube and by making a point of lingering on pictures of polar bears that popped up in our bedtime reading. I never presumed to know what she was experiencing or thinking—the emotional life of a child seems just as ungraspable as an animal's to me—but now, out on the tundra, I swore I saw in her some elemental experience of astonishment, the recognition of a familiar fiction materializing before her eyes. That's the experience Churchill's tourism industry is selling, after all, though we grown-ups can be more self-conscious about giving ourselves over to it.

At lunchtime, we parked at a bend in the coastline, and Robert and our driver set out sandwiches and soup. Someone noticed that a bear we'd previously seen in the distance was heading straight to us, hugging the shore. Maybe it was attracted to the smell of lunch. "She's going to come right alongside and check us out," Robert whispered to the kids as the animal crept within a dozen yards. Soon the polar bear was underneath his window, and he leaned over the edge and exhaled rhythmically, shooting muffled, punctuated breaths to attract the bear, wanting it to rise up and paw the side of the buggy as bears often do. Nearby, Sam blew in his hands and whispered, "You're beautiful, aren't you?"

We hadn't noticed somehow that this female was being followed by a smaller bear: a cub, or maybe a yearling. It was hard to tell. The adults aboard the buggy noted that the young bear looked unusually thin. It hunched by a back wheel. But when it got too close to the female, the larger bear huffed loudly and feigned a charge. The yearling turned and trotted away, spinning its head back toward the female as it ran. It stopped. Then, after a minute, it moved closer again.

This happened several times. Finally, the female flushed the younger bear for good and hauled off in the opposite direction. Suddenly, everyone on board was talking again, weaving a speculative story to explain what we'd seen. Maybe the female had weaned the yearling and was now chasing it away, as mother polar bears do. Or maybe the yearling wasn't actually her offspring, but an orphan—maybe a climate change orphan—and was trying to make this other female its surrogate mother. What we'd seen was either a brutal part of nature or a natural part of a brutal new trend. No one could say. It remained a drama with no narrative.

After lunch, Robert stood at the front of the buggy and asked people what their favorite moments of the day were so far.

"Having a bear make a connection with me," the youngest Vickery sister, Madison, said.

"And what is a connection?" Robert asked her.

"Having a bear look you in the eye, and you look at it," she said. She thought for a second. "There's no words to describe it."

"What about feelings?" Robert said. "What did people feel?"

Sam said, "Respect."

Robert seemed to like that. "That's a very important one," he told the students. He said that what he always feels is a kind of rattling insignificance: "The feeling that my problems are silly." And before I even realized it, he'd shuffled into the most signature of his signature riffs—the canned speech that would be written on that winter's Polar Bears International Christmas card to donors and which I'd already heard several times. It was a genuinely affecting monologue about how improbable and precious our Earth actually is—a paradise, "possibly the Garden of Eden itself"; a "little blue marble," suspended on the tail of the Milky Way. "If someone has given us a gift as magnificent as this," Robert would say, "I think we better take good care of it."

Frankly, it was a little unsettling how expertly Robert drilled his ambassadors, reinforcing his own talking points. It felt just slightly too manipulative, like an indoctrination—even to me, who believed wholeheartedly in the cause Polar Bears International was indoctrinating these kids into. In town, I'd listened to guys like Dennis the Bear Man criticize PBI for brainwashing the students they bring to Churchill. ("Some of these kids are babbling idiots by the time their week's over," Dennis said. "They're so neurotic and worried about saving the world. My God, I'm sure they must have nightmares.") And suddenly I could appreciate these men's disgust—not with Robert personally, but with the fact that it has come to this; that the experience of just being near these animals has become so loaded and solemn; that we've turned the polar bear into a psychic pack animal and heaped our shame, disquiet, and hope on its back.

"I love you," Robert told the kids, rounding out his blue marble speech. "I'm proud of you. Thanks for helping straighten out what we messed up."

THE EPISODE WITH THE FEMALE and the yearling was by far our buggy's closest and most dramatic encounter of the day. Isla missed it.

She'd gotten restless, no longer appeased by the chocolate chips we'd been dispensing like SeaWorld trainers reaching into their fanny packs for smelt. Then, quietly, mercifully, she had fallen asleep in Wandee's arms, her hand shoved down the neck of Wandee's shirt in lieu of holding a stuffed animal.

When the two bears approached, Wandee and I had agreed it was unwise to wake her. We were only four hours into at least an eight-hour day, and—to be honest—I never quite appreciated the magical appeal of staring into the eyes of a polar bear at close range anyway. It only reminded me of the artificiality of the situation in Churchill, the kind of reverse zoo that the tundra has become. It's the tourists, confined to a buggy, that depend on the polar bears to approach and interact with them. And if none do, many buggy drivers told me, it's the tourists who get bored or sometimes even insolent, griping on the ride home and stiffing their driver on tips. To me, the most moving part was always when a bear was done investigating the buggy and turned around and padded away, and how it kept getting smaller and smaller, swallowed by the endless negative space around it.

When Isla woke, I tried to explain what had happened outside. She pulled herself up to the window but could stare only at the vacant snow and ice. Sam was nice enough to show her the video he'd taken of the mother and yearling's stand-off; for Isla, in other words, it wound up being just another digital video clip. She watched with blank eyes. But when, on the screen of Sam's camera, the female huffed and the yearling trotted away—its head turned, its legs quickening into a momentary gambol—Isla seemed to experience that shock of recognition again. She looked at Sam, then at her mother and me, and said, very confidently, "Horses do that."

In retrospect, what I remember most about our day on the tundra is an overwhelming feeling of relief—how struck I was that Isla was actually enjoying herself. I'd been focused on what the experience might come to mean for her years or decades later, as though it were a financial investment or a vaccination. I'd forgotten that watching polar bears could be fun.

We'd been in Churchill together for a couple of days already, and so much of what Isla was encountering there was spectacularly new to her: walking on snow, sliding on ice, shooting down the tremendous polar-bear-shaped slide in the town's recreation center; the hilarious rustle that her snowsuit made when she ran back and forth down our hotel hallway. And she still seemed to have no saturation point—for her, newness never got old. In the last moments of our Tundra Buggy tour, we pulled alongside a small bear, slogging toward us from the middle distance. Isla leaned out of Wandee's arms and through the window. Then she turned to me and said, "Look, Daddy! Polar bear!" as though it were still surprising, as though the only two words she had for that thing still needed to be said out loud.

5.

THE LIFT

After we got home to San Francisco, I got an e-mail from Daniel J. Cox, an accomplished photographer who's been going to Churchill to shoot polar bears for twenty years and now volunteers for Polar Bears International as their in-house photographer every bear season.

Cox had been on a buggy with a tour for wildlife photographers when they came across a distressingly gaunt female polar bear fidgeting behind a snowbank. The temperature had finally dropped, and the wind was flaring, and this particular bear, without much fat left on her for insulation, was struggling to find a comfortable resting position. Her movements were erratic. Her shoulder blades arced above her head like a wishbone.

There was still no ice on Hudson Bay. The bears had come onto land that summer between late June and mid-July. They ultimately wouldn't get back on the ice until around December 4, having spent as long as 162 days on land, or forty more days than what used to be considered normal, and maybe far longer than many cubs would be equipped to last. It was shaping up to be the sort of potentially cata-clysmic off-year that the biologist Andrew Derocher had warned me about.

As Cox and his group watched, another, fitter polar bear ap-proached the withered female. The female labored to ratchet herself onto her feet and eventually scuttled out of the snowbank to charge it and scare it away. As she did, the twin cubs clustered under her body for warmth came into view for the first time.

One of the cubs rose to look around. Its face was lithe and fluffy; it looked more like a wolf pup than a bear. It started to lean backward strangely, like a tree bending in a wind. Then its face and mouth twitched. Then the twitching radiated into the rest of its body and intensified. The cub was seizing up—convulsing in what veterinari-ans later assessed to be the last phases of starvation. As it shook, the mother sat stoically beside it, swiveling her neck back and forth, scan-ning the tundra. The cub died shortly after. Its sibling died two days after that.

Cox was e-mailing me a short video of the episode that he'd filmed and posted online. The footage seemed to have the potential to go viral and become iconic, a concise and transfixing scene of the vio-lence of climate change, which is otherwise slow and abstract. Cox had put in a title card, explaining that, although science could never link the starvation of these two specific cubs to climate change, this was exactly what biologists expected to see more often. And he tricked out the scene with some acoustic guitar and cello music: This made it even more sorrowful without tipping into melodrama.

I'd read in many scientific papers about increased rates of starva-tion among cubs. But seeing a cub spasm and stagger through its last moments is a different experience, of course. Watching the video felt incriminating in the most paradoxical way: I felt unsettled by how much power our species is wielding on the planet, and I also felt powerless. In a way, the video represented the conclusion of the same story the teddy bear helped tell back in 1902: now, finally, society's reach has expanded all the way to the top of the world and demeaned even the most remote and mightiest bears.

Still, some people who watched the video asked a question that hadn't occurred to me: why hadn't Cox put down his camera and called in wildlife officials to feed the cubs, or even chucked out some of the tour group's hamburger meat? Some saw the video as an exploitative snuff film. It played right into the cynicism I'd encountered in Churchill about the motives of modern polar bear conservation—the suspicion that it's all empty PR. I'd come to realize that people in Churchill cared just as passionately about the species as anyone in Polar Bears International does. But if you refuse to accept the premise of climate change, as they did, then you also reject the idea that the survival of polar bears hinges on influencing some opaque emotional calculus of far-flung SUV drivers and politicians and the size of their carbon footprints. You believe, instead, that the survival of polar bears depends on *the survival of polar bears*—the physical welfare of the individual animals on the land. Dan Cox believed his footage could help polar bears. But you had to make a certain mental leap to see it that way. In the simplest terms, it was a video of him *not* helping polar bears.

Nasty comments proliferated online. Some people were angry. ("Shame on you, Mr. Cox, shame on you!") But some seemed only to feel betrayed, unsure what exactly they, as donors to Polar Bears International, were supporting. "Your agency is out there to help and protect these magnificent animals," one person wrote. (Cox had posted the video on his own Web site, but his affiliation with PBI is well-known.) One way for people to contribute to PBI is by "adopting a polar bear cub." And though it is only a "symbolic adoption"—for $100, you get an adoption certificate, a tote bag, and a gourmet white-chocolate polar bear—it was hard to reconcile that messaging with what they saw in the video.

Cox wrote a response, and leading bear biologists lined up to defend him, noting, for starters, that feeding bears is illegal, and that the cub was in such poor condition that it was likely to die regardless. But they most of all stressed that feeding individual bears would only put a Band-Aid on the problem of climate change—it missed the point. Robert Buchanan told me that the reactions he saw were indicative of a knee-jerk, "bunny-hugging" attitude that, frankly, he can't stand. Here were folks who burn a disproportionate share of the world's fossil fuels feeling self-righteous about polar bears. "They're killing polar bears from the comfort of their easy chairs," Robert told me. "Excuse my expression, but fuck 'em."

There's no accounting for the polar bear's magnetism. The bear seems to have evolved to wrench out human emotion as efficiently as it evolved to rip ringed seals out of the ice. Robert was betting the species' survival on that appeal. He believed that that emotional connection could be enlarged and channeled into action. But the empathy that Cox's video generated almost seemed to be too much—too raw and unwieldy to be channeled into anything.

To me, the starvation video hinted at a bigger problem: that, maybe, this approach to polar bear conservation was reaching its breaking point. The polar bear had been useful as a pure white, cuddly harbinger of horrible things—a victim of climate change at just the right remove from our own species to be palatable and approachable and inspire us to change. Until now, working to save polar bears and working to stop climate change have meant the same thing. But they're starting to diverge. The lives of the actual polar bears in Churchill—the most visible polar bears in the world—will become increasingly grim. Cubs will starve or be cannibalized, and a greater number of worn-down animals will crowd into the viewing area waiting for ice.

If we choose to help them survive, it will require a kind of narrow, hands-on management—like getting out there and feeding them meat—that does nothing to stall climate change. Meanwhile, conservationists will have to work hard and more inventively to spin the animals' worsening predicament in the same inspiring terms, without seeming dishonest or oblivious.

THE LIFT HAPPENED on a Friday afternoon. Not far from the Churchill airport, a crowd formed outside a huge Quonset hut left over from the military days. Yellow caution tape was strung along the roadside, and tourists gathered behind it, a hundred yards away, as more school buses pulled over to park, their flanks and windows discolored by a crust of dust and frost.

The Quonset hut, known as D-20, has been turned into what Manitoba's provincial conservation agency calls a "polar bear holding compound," though nearly every reporter who comes to town prefers the phrase "polar bear jail." It's the centerpiece of the government's Polar Bear Alert Program, a protocol for handling bears that wander into Churchill in the months before freeze-up. The program was put in place in the early eighties, partly in response to the mauling of Tommy Mutanen. Sightings of bears in town are reported to the agency through a hotline—675-BEAR—and bear patrol officers respond to haze it with pyrotechnics and noise-making shotgun shells or chase it back onto the tundra with their trucks.

Any bear that can't be chased away is drugged and transferred to one of the twenty-eight cells inside D-20. It's held there for about a month and not fed during that time, so it won't connect sneaking into Churchill with being rewarded with food. At the end of its sentence, if Hudson Bay still hasn't frozen over, the bear is drugged again, airlifted by helicopter, and released north of town, closer to where the ice first forms. The idea is to dissuade bears from entering town again. Once a bear becomes comfortable mingling with humans, there is usually no choice but to shoot it. There were nine polar bears inside D-20 that afternoon. If you put your head close to the building's corrugated side, you could occasionally hear them growling. Soon there would be eight.

A school bus pulled up around the back side of D-20. There was a kind of VIP section there, with Robert Buchanan and his people milling around in their matching blue Polar Bears International parkas. Martha Stewart's crew was there, too. Now off the bus came another group of dignitaries: the Canadian chapter of the World Wildlife Fund had arranged a trip to Churchill for some of its corporate partners, including a cohort of young executives from Coca-Cola, which has used polar bears in its commercials since the nineties. The World Wildlife Fund and PBI were actually paying for this afternoon's bear airlift from D-20. The government was grateful for the help in exchange for doing the lift at a certain time and allowing all these guests to watch the spectacle up close, and Martha's crew to film it. (PBI was also making its own educational film, which would be distributed to zoos and aquariums.) While we waited for the lift to start, I chatted with Steven Amstrup, Polar Bears International's senior scientist, and soon Martha's producer—the same raspy fellow from that afternoon on Buggy One—came to Amstrup with a question. The producer kept hearing people call polar bears the "world's largest land carnivore," and he wanted to know if that was true. Were they bigger than the Kodiak bears in Alaska?

Amstrup explained that polar bears are equal in size to Kodiak bears, which are the world's largest brown bears.

"So," the producer said, "can Martha say 'world's largest land carnivore'?"

"Well," Amstrup began again, "I'd like to correct that." Polar bears are actually classified as marine mammals, he said, since they spend most of their time on sea ice. But the polar bear certainly isn't the largest predatory marine mammal. That would be the orca.

The producer thought a second. He didn't know what to do with that information.

In the end, Amstrup and other PBI higher-ups would describe the segment that Martha produced as among the most solidly reported and properly messaged media pieces they'd collaborated on. (In an in-studio segment of the episode, while preparing a baked Alaska with the comedian Andy Samberg—the theme of the show was "Cold"—Martha even wore the blue Polar Bears International parka that the staff had given her, which, everyone had noticed, she declined to wear in Churchill.) Even now, outside D-20, there were signs that PBI and the television crew were starting to understand each other better.

"What do you want to call them?" the producer finally asked Amstrup.

Amstrup thought a second. "I like to say, 'Polar bears are the world's largest nonaquatic predators,'" he said, though he seemed to understand that it didn't exactly roll off the tongue.

The producer smiled good-naturedly. "We'll see if she'll say that," he said, and walked away.

IT BEGAN WITH A HELICOPTER landing right in front of us, a hundred feet from the D-20's open door. A man in a reflective vest hustled out and hitched a fluorescent orange cable connected to the chopper's undercarriage to the pile of black netting on the ground.

Then out of the Quonset hut came a small ATV, towing a plywood flatbed. The tranquilized polar bear was on it, flat on its belly, positioned to face backward, so that the ATV didn't blow exhaust in its face. Its fur was yellowing and crimped in places. Its huge muzzle was black with dirt. Two wildlife officers walked on either side of the bear in uniform, holding shotguns, like security guards or pallbearers.

When they reached the netting, the ATV driver climbed down. Then the three officers together lifted the plywood from its base. One cradled the polar bear's neck in his arms, as an EMT would, and they spilled the animal onto the netting. It was on its back now. Its left paw had landed across its chest, in the posture of a drunken uncle after Thanksgiving dinner.

Behind the yellow tape, every tourist was holding a camera. There was something ritualistic about the scene—the way no one watching or participating in it said a word, yet all the players knew their parts. It was a ceremony of our saving a polar bear, or at least going out of our way to coexist with it peacefully and not kill it when it encroached on our turf. One PBI volunteer later told me that she's cried at bear lifts in the past. The tears were partly because it's difficult to see an animal laid out and drugged and partly because she was just so happy. "I am saving a bear that, in earlier years, would have been shot," she explained. Here, really, was a metaphor for everything I'd seen in Churchill: the polar bear placed like a slack white prop in the center of a crowd that longed to do right by it, that was so impressed by the animal and cared so much that it could scarcely comprehend what to do with that concern.

Soon the helicopter's propeller churned again. The men came hurrying out of the noise toward the crowd. They were motioning at us with their arms. Their mouths were saying, "Stand back."

The chopper rose. The orange cable underneath it unwound and straightened. Then the edges of the netting began to lift. The furry shape inside folded in on itself, head and legs cradling toward the center. The bear contracted into a U. And then—impossibly—the entire package was off the ground, rising ten, then twenty-five feet in the air.

The helicopter soared over the nearby power lines and kept climbing. The bear twirled slightly like a tea bag beneath it.

Now it was quiet enough to hear the person next to you talk again. Robert Buchanan waved some of his staff, Martha Stewart and her crew, and a few lucky preselected executives from the World Wildlife Fund tour onto two other helicopters, waiting nearby. They would follow the bear and be on the ground when it was released. Some would take photos with the tranquilized animal's head in their laps and post the pictures on Facebook.

Those of us left on land could only look up, still reeling from the implausibility of what we'd just seen: the wild animal's ascension into a dirty white sky. It was all so obviously charged with meaning, but impossible to work out. I looked up and watched the image get tinier and fuzzier, until it was hard to know what it was anymore. Then, finally, the polar bear was gone.

PART TWO

BUTTERFLIES

6.

THE MIDDLE OF
A HAIRCUT

The Antioch Dunes National Wildlife Refuge sits at the very edge of Antioch, California, an economically haggard suburb an hour east of San Francisco. It's a narrow band of land at the anonymous, industrialized fringes of town, the kind of place not easily reached by any bus line, but around the corner from where a transit company parks its buses at night. The refuge is bordered by the San Joaquin River to the north and, across the road to the south, a sewage treatment plant. Its parking lot abuts a waste transfer station and a diner called the Red Caboose, a converted old Santa Fe Railroad train car that is popular with bikers.

At only sixty-seven acres, Antioch Dunes is among the very smallest of America's national wildlife refuges. You could walk straight from one end of it to the other in under an hour, if it weren't for the tremendous Georgia-Pacific Antioch Wallboard Plant, which, built before the refuge was established, still stands on a block of privately held land, splitting the refuge in two. The wallboard factory, which everyone calls the gypsum plant, is a daunting facility with its own water tower and container ship dock on the river. It makes drywall and emits a low-pitched thrum that you can hear when you're walking around the refuge. Sometimes gypsum dust—a chalky white powder—drifts over the dunes and settles on the leaves of plants.

I spent a lot of time at Antioch Dunes and grew to really like the place, but the fact is, it's hard to describe the refuge without sounding like you are insulting it. Even calling it "Antioch Dunes" can sound snide, since there have been virtually no recognizable sand dunes here for at least twenty or thirty years. Long before the federal government bought and protected the land in the 1980s, the sand that had piled up spectacularly was gradually trucked away to make bricks and roads and at least one horse-racing track, until it was virtually all gone— dug down to near the water table in places—and a rowdy garden of shrubby, nonnative weeds and trees soon sunk in its roots. In an essay titled "In Memoriam: The Antioch Dunes," a botanist accused the landowners who sold that sand of converting the area's "permanently uncommon value into transient cash" and forever leaving "the people of California that much more bankrupt in soul and spirit." He was writing in 1969, arguably before the place went most dramatically downhill.

Recently, one afternoon in late August, a Fish and Wildlife Service employee named Louis Terrazas was showing a group of volunteers a large aerial photo of Antioch Dunes—giving us the lay of the land. We had assembled on the smaller of the two halves of the refuge, east of the gypsum plant, to count butterflies. Louis brought cookies and sunscreen for everyone. As we helped ourselves, he leaned the photo against the torso of a squat, slightly thuggish guy in camouflage shorts named Steve, using him like an easel. When I'd asked Steve what brought him out to count butterflies, he told me, "I'm here because I was a bad, bad boy." It was a long story, which I couldn't really follow, having to do with Steve's tendency to run stop signs while doing a paper route at three in the morning, but the upshot was that a judge finally got tired of slapping Steve on the wrist and sentenced him to some whopping number of community service hours. He was handed a list of possible jobs. Louis was the only guy on the list who called him back.

We were here to count a specific kind of butterfly: the Lange's metalmark, a little-known but very critically endangered species. Antioch Dunes is the only place on Earth where the Lange's metalmark lives. (There are also two endangered flowering plants here, the Contra Costa wallflower and the Antioch Dunes evening primrose—the refuge is the only one in the nation set aside for endangered plants and insects.) As a baseline for the butterfly's conservation, the government needs to establish each year's "peak count," or the highest number of Lange's spotted on a single afternoon. In the nineties, peak counts reached into the thousands. But the species was flirting with extinction, Louis now told us. In the summer of 2006, the Fish and Wildlife Service had been shocked to find that the population had suddenly crashed. Peak count was only forty-five that year, which is to say that all the Lange's metalmark butterflies known to exist on Earth one afternoon could have fit inside a French press. "So . . . I wouldn't see them in my backyard?" said an older woman in a Puerta Vallarta baseball cap. She'd adjusted her voice mid-sentence so that, by the end, she was clearly answering, more than asking, a question.

There were sixteen of us volunteers—a few older couples, a garrulous oil industry grunt at Chevron, a college student with a Day-Glo tiger tattoo on her back who'd heard about the butterfly count on Craigslist. Except for Steve, the traffic violator, we were all here because we wanted to be—for the benefit of the butterfly, or our own benefit, or some inscrutable intertwinement of the two. Butterflies occupy a special place in our imaginative wildernesses, transcending their status as bugs. They don't sting, bite, buzz in your ear, or scamper across your kitchen floor. If you woke up to find one had alighted on your nose, you'd lie perfectly still, puzzling out whatever beneficent message the cosmos must be communicating to you—whereas you wouldn't do this if you woke up with a banana slug on your nose, or a cockroach. We see butterflies as delicate, uncorrupted—which is probably why we're so keen to paint them on our daughters' cheeks at birthday parties or stitch them in glittery thread on their pajamas. We had to invent unicorns and fairies to keep little girls company. But we let the butterflies in, too, just as nature made them.

In that sense, the story of the Lange's first sounded to me like a melancholy children's book. A fragile, beautiful butterfly had been hemmed in on all sides by filth and modernity until its home—fenced off, closed to the public—faded into a kind of forgotten badland. Stolen cars have turned up at Antioch Dunes. Once, Louis told me, a biologist found the body and head of a dead pig in two separate black garbage bags, dumped at the front gate. And a short drive away is the house that belonged to Phillip Garrido, who kidnapped eleven-year-old Jaycee Dugard in 1991, held her in a warren of outbuildings in his backyard for eighteen years, and fathered two daughters by her. In our imaginations, butterfly habitat is always a pretty place on a beautiful day. But somehow the Lange's metalmark had persisted here long enough to attract some human attention—not much, but enough, in our age of conservation reliance, to change everything.

The Antioch Dunes came apart slowly; it took the entirety of the twentieth century for them to devolve into what they are now. The butterfly was tossed around on the surface of that confusion. There were, I'd find, people tossed around, too—people who loved the place, and other neglected landscapes like it, and tried to counteract the entropy taking hold. I found their stories rising and falling in the history of the Antioch Dunes, almost cyclically: Each generation of idealists was running after butterflies, trying to save them but never quite catching up, until, having watched so much nature deteriorate in their lifetimes, they finally buckled over, jaded. And just as they did, without fail, the next generation of idealists would appear, obliviously hitting their stride.

I wanted to trace the changes at this one specific track of land across all that time. But looking so closely at an insignificant-seeming bug, at an insignificant-seeming place, eventually drew me into deep uncertainty about so much else. It turns out that the Lange's metalmark flies in a confounding and counterintuitive wilderness, where some of our most comforting ideas about nature unravel. I kept chasing the butterfly, wherever it led me. And before I knew it, I was all the way back at conservation's first principles, faced with petrifying questions like, what exactly are we preserving, and why—questions worth asking, even if they can't be answered.

LOUIS PASSED AROUND some laminated photos of the Lange's metalmark so we would recognize it in the field. It was orange and black and looked to me like a smaller Monarch butterfly, but probably only in the way that all unfamiliar meat tastes like chicken to the uninitiated.

This was the middle of the Lange's flight period, he explained—the few weeks every summer when the butterflies emerge from their cocoons and zip from place to place, mating and laying eggs. It lasts about a month, until they've all been picked off by dragonflies or succumbed to old age. (Butterflies live hard and fast.)

It was up to us volunteers to form a long line and pace every transect of the dunes with handheld clicker counters, trying to spot the butterflies, one by one, as they quivered through the air. The trick,

Louis said, was to stay in line; anyone who got too far ahead might spook a Lange's and flush it out of the plants before it could be positively ID'd. Also, we had to keep looking behind us for butterflies scattering in our wake. It sounded hard. "You ready?" Louis shouted.

He gave the signal, and we began creeping through the brush. We moved in almost perfect lockstep for a few paces. Then someone saw something and shouted. Someone else yelled, "No, I think it's a buckeye," noting that the buckeye butterfly, also pictured in Louis's mug shots, was bigger than the Lange's.

"We got one over here!" I heard Louis yell from the far end of the line. He wanted everyone to keep walking in formation, but our heads were turned now, and there was an immediate and unmistakable drifting to the right, toward Louis and the butterfly. Then someone on the opposite end of the line yelled, "There's another one right here!" and some of us started drifting that way, too.

"Is that a second one?" one older women asked.

"There's one right here, too!" a man hollered.

"Rock on!" someone said.

"Oh, here's a third one!" a woman shouted, though it was unclear whether one of the first butterflies had merely jittered through her line of sight.

There was a fifth and sixth sighting, maybe more. Fingers were firing on the clicker counters. "We've got another one here!" Louis yelled.

It was a monstrously eventful and confusing ten seconds. And in that pandemonium, it was immediately clear just how unscientific this process was going to be. The very baseline understanding of the species' health was being provided by us, a bunch of civilians, who had only just been shown a photo of the bug a moment ago. And yet this is a common situation. As the budget for protecting endangered species and managing wildlife has stayed relatively stagnant, but the workload has exploded, more of that work has fallen to a standing army of curious and often retired volunteers—citizen scientists whom Princeton ecologist David Wilcove has compared to volunteer fire-fighters. In Maine, they count moose and frogs. In Ohio, they snatch Lake Erie water snakes out of the water and measure them.

When the excitement was over, Louis went around to the witnesses individually to reconstruct what had happened. He felt confident that the tally ought to be capped at five. Then he called us back to take a good look at the first Lange's metalmark, which was sitting obligingly still on the stalk of a buckwheat plant. We huddled around it. The girl with the tiger tattoo took a picture with her iPhone.

I squatted and looked at the butterfly for a long time. It was the size of a quarter. The wings were rimmed in black with white speck-les, then gave way to sunbursts of deep orange. I'd seen lots of photos of the species before that afternoon, but the butterfly was always blown up and perfectly centered in the shot. Looking at it now for the first time in the wild—seeing it as a tiny blotch on a big leaf, with so much air and space and civilization around it—brought a deflating new sense of scale. The bug seemed vulnerable to the point of helpless-ness. You wanted somehow to zoom in, to make it feel important and central again—a worthy protagonist of the bizarre, generations-long saga that's played out at Antioch Dunes on its behalf.

You wanted to make the butterfly look big again. And this could be why one of the older women in our group had taken to her knees only a couple of inches from the leaf and was now examining the butterfly through binoculars.

ONCE, THE ANTIOCH DUNES WERE what ecologists call a dis-turbance ecosystem. The landscape was in motion—slow motion. Wind-driven sand gradually piled into new dunes, and older dunes periodically collapsed. Certain plants thrived in that unruly environment, while the seeds of newcomers were never able to get a foothold in the shifting sand.

One native that's well adapted to this cycle is a tall, spindly plant with muted white flowers called "naked stem buckwheat." The buckwheat is the Lange's metalmark's host plant. Every butterfly lays its eggs on a particular host plant, and whereas some species of butterflies are promiscuous, laying on a variety of plants, others, like the Lange's, are committed to a single one. Strictly speaking, the host plant is the butterfly's habitat—the platform it needs to survive, like sea ice for polar bears. Butterflies will thrive in even the most totally run-down-looking or artificial landscape if they find enough host plant there. In Miami, conservationists have expanded the range of a species called the Atala by planting its host plant in highway medians and attracting the butterfly there, into the middle of traffic.

If you watched a time-lapse film of Antioch Dunes, you'd see, as dunes formed and fell asynchronously around the property over the decades, different plant species mushrooming and then dying back on top of the dunes in a predictable succession. The buckwheat is part of that succession. Every summer, some metalmark eggs laid on stands of buckwheat the previous year will turn into butterflies and take flight. Others will have had their plants senesce or the dunes blown out from under them. As certain encampments of butterflies dwindle, butterflies from other, nearby stands of buckwheat skitter in to lay eggs there and supplement them—what's known as the "rescue effect." Or they pioneer new colonies elsewhere, as new patches of buckwheat mature into good habitat. Around the dunes, individual colonies of Lange's metalmarks will grow and die out chaotically. It's not uncommon for the total number of butterflies in this kind of "metapopulation" to spike and dip dramatically from year to year. But in the aggregate, the meta-population survives; there's enough of a cushion so it can recover from any losses.

But that cycle at Antioch Dunes has now ground to a stop. With the sand largely gone, dunes have stopped rising and collapsing. And without that disturbance, invasive plants could settle into the stable ground and overpower the native ones. The stands of buckwheat have become fewer—and farther between. An individual Lange's metalmark may never venture more than a thousand feet from the buckwheat plant it hatches on—the range within which it can lay its eggs is extremely limited. So the distance between certain butterfly colonies can easily become insurmountable, cutting off the rescue effect. One big meta-population fractures into smaller, isolated islands, each clinging to its buckwheat and surrounded by an uncrossable sea of weeds.

In recent years, the most recklessly spreading weed has been a scraggly, purple-flowered legume with a sinister-sounding name to boot. It's called hairy vetch. The insidiousness of hairy vetch, and the many levels of damage it does, can't easily be summed up. (In some cases, the vetch physically wraps itself around buckwheat plants and steals their sun.) The plant is Louis Terrazas's nemesis. As the primary person assigned by the Fish and Wildlife Service to manage the habitat at the dunes, he spends his year in one long counterattack against the vetch. (Theoretically, he can mow, whack, or blast pesticides at it from a backpack sprayer, but he has to be careful not to harm any stands of buckwheat in the process, since they may have Lange's larvae on them, or any of the federally endangered plants at the refuge, which he painstakingly marks in advance with fluorescent orange flags. Truthfully, a lot of Louis's job comes down to yanking vetch out of the ground by hand.) When I first visited, a few months before the butterfly count, Louis was throwing everything he had at the vetch, trying, alongside the occasional community service worker, to whip the landscape into shape before the butterflies started breaking free from their cocoons later that summer. But the vetch was still everywhere. "This place eats our lunch," Louis confessed. He pointed out areas that he'd cleared that spring but which the vetch was already recolonizing. Elsewhere, other nonnative plants, like ripgut brome and yellow star thistle, had crept in to fill the niche that Louis had torn the vetch from. "Right now," he said, "we're kind of like in the middle of a haircut. Like, when you look in the mirror and say, 'Man, this isn't looking so good.'"

But in a sense that haircut will never end. It's easy to call the vetch a weed and rip it out of the ground to give the buckwheat some space, but the place has changed so completely that, on some level, the weeds now belong at Antioch Dunes more than the native plants do. They are heartier, thriving—better adapted to the new ecosystem that human beings didn't realize they were creating by taking away all that sand. Keeping the Lange's metalmark at Antioch Dunes means keeping the naked stem buckwheat there. But that now means keeping someone like Louis Terrazas there, too, slogging around with a weed whacker. If the butterfly is going to survive, we have to simulate the disturbance in the ecosystem now—we have to be the wind. As another Fish and Wildlife employee told me at the dunes one afternoon, "This place will never run on its own."

There'd been wariness of conservation reliance like this looming in the back-and-forth over whether Dan Cox should have fed those starving polar bear cubs in his video. Lost in the ignorance and outrage were a few sensible people who claimed to understand the problem of climate change perfectly, and to understand that feeding individual polar bears wasn't going to remedy the larger, more horrible situation. But they still wondered if we should be feeding the Hudson Bay bears anyway—because, like it or not, that's what might be necessary now to keep those animals in the world. Conservationists cast any long-term effort to feed polar bears as logistically problematic, prohibitively expensive, and dangerous. But it's also philosophically ugly. Making polar bears dependent on us for their very survival in such a hands-on way can feel like just as much of a defeat as letting them die out. It would mean conceding that their ecosystem is irreparably broken, and that we have to be responsible for them in perpetuity, not just step in temporarily to save them. It feels too much like playing God—even if, arguably, that's exactly what we've become. After all, we're the ones who upset so many ecosystems in the first place—we override the natural course of evolution when we endanger species, too, not just when we try to save them.

Still, once you purposefully cross that line, it's not clear where you would draw a new one. Some people write to polar bear biologists suggesting that we feed polar bears; others imagine melting down plastic soda bottles to build motorized rafts, so that the bears can float around the Arctic as the ice disappears. Once you go hands-on, in other words, you have to decide when you're going to take your hands off. As J. Michael Scott, the conservation biologist who helped coin the term "conservation reliance," put it to me, "I could keep polar bears alive in San Diego if I really wanted to."

And yet here at Antioch, on behalf of a tiny butterfly that no one's ever heard of, on a refuge where no one goes, America seems to have quietly blown past that threshold a long time ago without ever really considering these questions. With the Lange's, as with other endangered species, we've gone all in, dramatically manipulating its ecosystem to promote its survival—sending Louis out there to groom the dunes to accommodate the butterfly's predilections and plant new arrangements of its favorite plant, as though he were the landscaper of some aristocrat's country estate.

Which is not to say that it's working. That afternoon in August, we were embarking on the sixth volunteer butterfly count of the summer at Antioch Dunes. The surveys had begun three weeks ago, but no one had spotted any butterflies until a couple of days earlier, when a single Lange's was seen on the opposite side of the gypsum plant. That is, when we started that morning, peak count was one.

WE DIDN'T SEE ANY more Lange's after that initial burst of five. One day the following week, the count hit twenty-eight. But that would prove to be peak count for the year. It was a new all-time low, and sent those working on the recovery scrambling.

One of the last transects we surveyed that afternoon was on some of the refuge's highest ground, a weedy dune with an old utility tower on it, once used to anchor power lines before running them across the river. The lines now ran through a second, newer tower, across a bowl-shaped valley, on an equally high hill.

Louis positioned us in a line abutting the bottom of the tower. Some of our waists were actually touching the rusted crossbar at its base. We had to survey the land under the tower, which meant that, on Louis's signal, some of us would be climbing over that steel bar into the tower's skeletal interior, then clambering between the metal cables and support structures and out the other side. If anyone had managed to keep up the fantasy that we were spending the day in a pristine wilderness, this conclusively killed it. Just as Louis was about to say go, one of the volunteers noticed a gap in our line. His name was Liam O'Brien, and he'd been fighting back a measure of cynicism all day. "I'll go over here in the gap," he shouted. "God forbid any metal-marks are *there*."

Liam is an energetic and knowledgeable butterfly lover who, when we'd first gone around a circle introducing ourselves that afternoon, elicited speechlessness and some actual gasps by relaying that he'd been out to count Lange's in 1997, and that they'd counted more than twenty-two hundred of them that day. He is forty-eight and lanky, with styled, whitening hair. He tends to move and speak in choppy, purposeful bursts. A moment earlier, he'd whirled around, tracking a fast-moving flickering in the air for as long as he could with his finger, and shouted, "That looks like a dogface! That's a California dogface!" hoping to give everyone the opportunity to see California's state butterfly. At one point, as we shuffled between transects, Liam whispered to me, "I'm not the annoying butterfly guy, am I?" I started to tell him that I didn't think he was; he was offering his little factoids in a helpful, totally egoless way. But something caught his eye, and he shot off in another direction. "Heliotrope!" I heard him say. "Great, great plant."

Liam and I had driven to the dunes together that day. I'd met him earlier in the summer in San Francisco and would occasionally seek him out at the café where he has coffee every morning, to bounce butterfly questions off him. He made his living with butterflies. He led butterfly walking tours in San Francisco every spring and summer. He painted butterflies to illustrate trail signs for local parks. He'd started a small, neighborhood-wide recovery of a butterfly called the green hairstreak and, despite being mostly self-taught, had also spearheaded the government's reintroduction of the endangered Mission blue to a hilltop in San Francisco called Twin Peaks.

That afternoon's metalmark count had left Liam in an introspective mood. Driving home, he told me that, as much as he loved butterflies, he wondered if the whole enterprise to save the Lange's was becoming a little foolish. "The tsunami of change that's going on at that place, with the nonnative weeds—you want to know, is this just an exercise in futility? What is it going to take to put these pieces of a puzzle back together when the puzzle itself has already changed?" He wondered about his own butterfly conservation projects in the city, too: whether he was wasting his time; whether, via some indiscernible chain of causes and effects, he might even be doing more harm to the environment than good. Totally possible, Liam said. There was no way of knowing. But in the end, he told me, "I just want to be part of a generation that tries."

7.

SHIFTING BASELINES

Butterflies swarmed the center of Liam O'Brien's life abruptly, fifteen years before I met him. Before that, he'd spent a decade as a professional actor. He mostly appeared onstage, doing Shakespeare and musicals, but he got a short-lived break in film in 1990, when he was cast alongside David Cassidy in a hokey sci-fi comedy called *Spirit of '76*. (Liam played an evil geek named Rodney Snodgrass.)

Six years later, he was living in San Francisco, working as an understudy in a production of *Angels in America*. And that was when it happened: a tether snapped tight between Liam O'Brien and butterflies. One day, he saw a butterfly with electric yellow and smudgy black wings landing in the garden outside his bedroom window. It was a western tiger swallowtail, a native of San Francisco. Liam had always kept a notebook of illustrations—a kind of visual journal—and, the next thing he knew, he was out in the garden with his pens and watercolors, capturing the swallowtail on paper. Soon he was touring around California in his Econoline van, painting and drawing as many of the state's butterfly species as he could find.

In 1998, Liam tested positive for HIV. He'd been losing his motivation to compete for acting gigs, and, in a way, the virus gave him permission to focus on the thing that made him happiest: butterflies. He set out to learn more, to collect names and explanations for what he was seeing. He went to annual butterfly counts and started hanging around some of the most respected butterfly scientists, or lepidopterists, in the Bay Area—a klatch of stoic, sometimes crotchety old men (they were mainly men) accustomed, since childhood, to chasing butterflies through woods, bogs, and canyons by themselves. Liam, on the other hand, is a snarky, spirited gay man with a big booming voice and no scientific background. But somehow he managed to win that crowd over. He found it ironic: We associate butterflies with feminine, gentle things. People use the terms "butterfly chaser" and "mariposa" as slang for gay men. "But I just happen to be gay. I go to the Lepidopterists' Society's annual meetings, and I've never seen such a collection of shabby straight men in my life."

One of Liam's mentors gave him a piece of advice: "Learn where you live"—dig deep, and study what's around you. So, in 2007, Liam decided to conduct his own exhaustive survey of San Francisco's butterflies, trying to see which butterfly species historically found within the city limits survived there. He spent more than two hundred days in the field that year, walking the defunct naval yard, the oceanfront scrub, and the weedy hillsides between posh Victorian houses, noticing which butterflies flew where. He was looking around, getting to know his neighbors.

Butterfly-wise, the San Francisco Bay Area happens to be a national treasure. The profusion of butterfly species in the region is arguably unparalleled in the United States; there were as many as fifty-seven before the Gold Rush. This diversity is a function of the severe changes in climate across the region, from dank, wet, and foggy to sunny and hot; even temperatures in different neighborhoods of San Francisco can differ by twenty-five degrees on a given day. This patchwork of microclimates creates something akin to the Galápagos Islands for butterflies, with different species and subspecies attuned to each area. In 1849, a French lawyer came to the region in search of gold but wound up chasing butterflies instead. He discovered about forty-three new species, eight from San Francisco proper.

The city became a hot spot for lepidopterists in the late nineteenth century. It was a time before biology and natural history had professionalized as fields, when a lawyer with a good eye and durable walking shoes could make valuable contributions. The most accomplished lepidopterists of the era were weekend warriors. One was a professional stage actor, like Liam. Another, Hans Hermann Behr, a physician, was once described as "always in danger of falling into forgetfulness on professional subjects when he caught sight of a butterfly he ardently wanted." Butterfly collecting wasn't a dignified hobby. It was seen as childish and useless—dorky. Although the stigma wasn't ever erased, a more glamorous, swashbuckling counterimage of the collector did momentarily emerge. It was best embodied by one of Dr. Behr's pupils, a San Francisco police officer named James Cottle.

Cottle was a huge, barrel-chested patrolman whose powerful fist, it was said, was more effective than his billy club. He'd performed heroically during the earthquake and fire of 1906 and once infiltrated a masquerade ball dressed as the duke of Wellington to catch jewel thieves. In a 1910 profile headlined "By Day He Catches Burglars; By Night He Catches Bugs," the *San Francisco Call* presented the cop as a brawny rejoinder to the image of entomologists as "blue spectacled old men, with long hair and pasty faces peering into slimy pools." Cottle himself had once felt that way about butterfly lovers. But he now understood that collecting butterflies was macho, athletic stuff.

He claimed to have once trudged across thirty-eight miles of problematic wilderness in pursuit of a single butterfly, the way hunters tracked bears. (Elsewhere, there were stories of men dangling hundreds of feet down the side of a cliff to capture a particular species of butterfly.) Cottle called butterfly collecting "the healthiest life in the world and the greatest sport I know anything about." A few months spent charging alongside him would cure tuberculosis, he said—guaranteed.

Boosters played up the monetary value of rare specimens. In the Bay Area especially, all the breathy fantasies of the Gold Rush were recast around butterflies. There were fortunes to be made in these "flying nuggets" of gold—all you had to do was reach up and swing a net. With new laws and conservation organizations like the Audubon Society making other popular naturalist hobbies like bird and bird-egg collecting problematic, some of that attention shifted to butterflies. And while the nature fakers were sympathetically anthropomorphizing wolves and bears, it was tough to feel bad about killing butterflies, particularly since they disappeared en masse at the end of their seasonal flight periods anyway. They were like the leaves, which painlessly dropped from the trees. As one how-to guide for collectors put it: "Their lives are so brief, what can it matter?"

There are no cases in which overcollecting was the major cause of an American butterfly's extinction, the way overhunting and egg collecting clearly wiped out birds like the great auk. Typically, butterfly species have been imperiled by the same, sometimes interrelated forces that threaten the Lange's metalmark today: habitat loss or habitat fragmentation, and invasive species. As early as 1875, in fact, Hans Hermann Behr, the absentminded physician, believed that the Xerces blue, a brilliantly blue-winged butterfly found only in San Francisco, had been driven extinct by the city's expansion. The Xerces was a sand dune species, but its habitat was changing from open sand to residential neighborhoods in Behr's lifetime. "The locality where it used to be found," he complained, "is converted into building lots." And the only insects that might survive in such a place were the "louse and flea." For Behr, it was only one example of how quickly the natural beauty of the Bay Area was disintegrating.

James Cottle was barely a teenager when Behr lodged these complaints; the landscape that Cottle's mentor now saw as spoiled was the only one that Cottle had ever known. Nevertheless, by 1928, an aging Cottle felt the same way Behr had at the end of *his* life, like he'd watched the beautiful terrain of his youth crumble into an ecological ruin. All his favorite butterfly collecting spots in San Francisco had been "erased forever by the city's growth," Cottle wrote—"rendered sterile and worthless . . . destroyed, defiled, eradicated." He saw scarcely any turf left, and focused instead on one day joining his mentor Dr. Behr in heaven. Cottle pictured Behr and all the other old collectors who'd preceded him waiting for him in the afterlife, staking out a tract of prime butterfly habitat "with nets enough to go around."

Liam O'Brien was doing his survey of San Francisco nearly eighty years later, in 2007. He counted thirty-two species and subspecies of butterflies, about two-thirds of what had been there when Cottle started collecting in the city. "In San Francisco," Liam told me, "we're known for what we've lost." The Xerces blue in particular stands as a chastising symbol—the nation's major invertebrate conservation nonprofit, the Xerces Society, is named for it. It ultimately became the first American butterfly known to be wiped out by humans, though it hung on in San Francisco many decades longer than expected. The butterfly had been presumed extinct for years when, in 1941, two young entomologists happened upon a small number of Xerces near a creek in San Francisco's Presidio. Ecstatic to see the butterfly still alive, they netted and killed large numbers of specimens to trade with their friends. It was the last time the butterfly was ever seen.

"I always thought there would be more," one of the entomologists told a reporter near the end of his life. "I was wrong." His name was William Harry Lange. A few years before the Xerces incident, he'd discovered a new kind of metalmark at Antioch Dunes.

HARRY LANGE, as he was known, was another native San Franciscan butterfly buff. He was born in 1912 and grew up collecting butterflies before school with an obsessive glee—at the same time, mind you, that James Cottle was on his way out, growing convinced that the city had been trashed. In fact, Lange went to high school within a couple of miles of several other budding entomologists, all of whom were so enraptured by the insects they found while tromping through their city neighborhoods that they pursued the science in college and quickly made names for themselves in the field. The 1930s and early 1940s are now considered a golden era for the science in the Bay Area. When Harry Lange netted his first Lange's metalmark at Antioch Dunes in 1933, he was merely one in an eager army of young men out there, scouring new and mesmerizing critters from the sand.

During the Depression, it was difficult for entomologists to travel to far-flung places for research. The Antioch Dunes emerged as an alternative study site for entomologists at the universities in Berkeley and Davis—and an endlessly fascinating one. At one time, there may have been as many as four or five thousand insect species living there, sewn symbiotically into the ecosystem. It couldn't have been a more convenient place to study, either: at the edge of the dunes was a bar called the Little Corral, whose owner allowed the men to park their Model A's there as long as they bought a beer first, and after a long day of work they could take swims or go fishing in the river. Consequently, probably no place in North America of such a small size has been scrutinized by entomologists so thoroughly, over so many years.

By Labor Day weekend 1954, the dunes were so well-known that, when a regional chapter of the Lepidopterists' Society had its annual meeting in San Francisco, organizing a field trip to Antioch was a no-brainer. The lepidopterists were coming to see the Lange's metalmark, and everyone managed to catch one and take it home. Still, Jerry Powell, an undergraduate at Berkeley at the time, later remembered being surprised that the generation just ahead of him—Harry Lange's generation—had written off the dunes by then. To them, the place was destroyed.

There had already been decades of intermittent sand mining by that time. It began in earnest after the San Francisco earthquake of 1906, when bricks were needed to rebuild the city. But it was especially vigorous when Jerry Powell first visited. At times, two separate companies were hauling sand from five different areas of the dunes at once, shipping it out by truck, river barge, or train. Spurs ran from the nearby railway to the area between the two power line towers, and sand was being shoveled out of that area most intensively, straight onto railcars. A small depression was just starting to be noticeable there: the beginnings of the bowl-shaped valley that I'd find gaping between the power line towers on the property now.

There had also been lots of industrial development around the dunes in the years after World War II, though, according to Powell, the last straw for Harry Lange's generation would be the construction of the gypsum plant in 1956. Not only did it sprout up in the middle of the habitat, splitting the dunes in half, but it also cast white dust over the place. It felt—in a very visceral way—like an atrocity.

It was all a matter of age and perspective, of course—all part of the same cycle of disillusionment that had been going on in San Francisco. The Lepidopterists' Society trip was the first time Jerry Powell had ever been to the dunes, and he and the other students were just as excited by the place as Harry Lange and his contemporaries had been twenty years earlier. Just as Cottle loved the city that Behr wrote off, and Lange loved the city that Cottle wrote off, Powell now loved the dunes that Lange's generation wrote off. Powell didn't see a landscape sapped of life. He saw a "terrific variety" of insects. And he was sufficiently captivated to study the ecosystem on and off for decades as a professor at Berkeley.

Two decades later, though, by the mid-seventies, development had winnowed down the dunes even more dramatically. Weeds were rampant, and another recent surge of sand mining had eaten farther into the property than ever before. But Powell was still intrigued. In 1976, he began what could have been a culmination of his research at Antioch Dunes: a study to plot the decline or disappearance of insect species at the dunes since the early 1930s, to describe how that web of life broke down.

Because the dunes had been visited so regularly by entomologists, Powell assumed that the specimens they collected in any given era were a good reflection of the insect life that existed there then. So he and his students searched through museums, where specimens were labeled with information about where and when they were caught, and tallied all the samples brought home from Antioch in two seven-year periods, a decade apart, starting in the 1930s. Meanwhile, Powell made his own trips to the dunes for the next seven years, collecting everything he could. Now he had samples from three seven-year time periods—a chronological catalog of what was netted at the Antioch Dunes on some six hundred different days between 1933 and 1983. By looking at when species were last collected, he could narrow down when they vanished from the ecosystem.

To make the study manageable, Powell had decided at the outset to look at only a fraction of the insect species known to exist at the dunes—376 of them, which still took him twenty-five pages to list. Powell concluded that 243 of those 376 species were now gone. All manner of wasps, beeflies, beetles, robberflies, and velvet ants had disappeared. Among them were four species endemic to the dunes—species, like the Lange's, that can't be found anywhere else on Earth. That is, they were not just missing from Antioch; they were now globally extinct.

But there was a serious wrinkle. Powell expected to show the number of species declining during those fifty years. Instead, his data showed the opposite. Even though many species disappeared, the total number of species being collected actually rose over time. Against all odds, biodiversity appeared to be *increasing* at Antioch Dunes.

He soon realized this was an illusion. Powell thought about entomologists' nets: how, in his lifetime, their mesh had gotten finer and finer as scientists became interested in catching insects of smaller size. In the thirties, California was full of insects that hadn't been named—large, alluring, and conspicuous bugs that leapt out at you from the landscape. The boys of Lange's generation concentrated on the butterflies and dragonflies and wasps—the charismatic megafauna of the invertebrate world. As those larger insects lost their novelty, or their population declined and they became harder to find, the next generation was driven to root around for smaller, more obscure things. And so, in turn, was the generation after it.

Over the years, the gaze of entomologists gradually magnified, each generation scrutinizing what the previous one hadn't bothered with or noticed. By the time Powell was surveying the dunes in the late seventies and early eighties, the insects he was bringing home included the minuscule and the nocturnal—because that's what a scientist of his generation was accustomed to collecting, and what was left to be caught.

The biodiversity of the dunes hadn't expanded. But people's perception of it had.

THE PHENOMENON THAT Powell stumbled onto has a name: shifting baselines syndrome. The term was coined in 1995 by a fisheries scientist named Daniel Pauly. Pauly recognized that global fish populations have been slowly collapsing, and though scientists weren't blind to that damage, their vision was too narrow and subjective to take in its full extent. Every generation of scientist accepts the oceans as it inherits them, Pauly argued. Overfishing may eat away at fish stocks, or even drive species extinct. But when the next generation of scientists start their careers, they don't see the oceans as depleted; that depleted condition becomes *their* baseline, against which they'll measure any subsequent losses in *their* lifetimes.

Because of this, a comprehensive picture of the changes happening across generations never truly comes into focus. Scientists are concentrating on only part of a line graph that is, in fact, much longer and more steeply plunging. (We now know, for example, that between 1850 and 2005 overfishing reduced the cod population in the northwestern Atlantic by 92 percent.) As we began to fish bigger species like cod into scarcity, we transitioned to eating smaller ones, like monkfish. As Pauly puts it, humans are blindly fishing their way down the marine food web—not any differently from how entomologists blindly moved down the web of insects at Antioch Dunes, with Jerry Powell fascinated by the tiny bugs that his predecessors let pass through the mesh of their nets. When Pauly introduced the idea of shifting baselines syndrome in the nineties, he often joked to the press that kids might soon be enjoying jellyfish salad sandwiches, instead of tuna. These days, he points out that there actually is a commercial jellyfish fishing industry up and running in Asia and the American Southeast.

Shifting baselines syndrome, then, is only the scientific manifestation of a broader problem affecting all people: what the psychologist

Peter H. Kahn Jr. has named "environmental generational amnesia." All of us adopt the natural world we encounter in childhood as our psychological baseline—an expectation of how things should be—and gauge the changes we see against that norm. This explains why the children Kahn has interviewed in terribly polluted neighborhoods in Houston don't believe their neighborhoods are polluted, and why Kahn's daughter thinks the woods around their family cabin in Northern California are beautiful and pristine, while Kahn can't get over no longer hearing the calls of owls. It's also why Hans Hermann Behr and James Cottle—and even Liam O'Brien today—could all spend their lives equally entranced by a San Francisco butterflyscape that only got progressively poorer. As Kahn puts it, "We don't know what we are missing."

Acknowledging the problem of shifting baselines syndrome, like truly acknowledging the enormity of climate change, can be profoundly disruptive and discouraging. It begs the question of what baseline biologists should be measuring wildlife populations against in the first place. It also can leave us, the public, unsure how to feel about conservation's supposedly feel-good success stories. In 1973, when the bald eagle was placed on the endangered species list, there were believed to be only 417 nesting pairs of birds left in the lower forty-eight states. In 2007, Fish and Wildlife triumphantly delisted the eagle, having by then built that population up to ten thousand nesting pairs. But the agency also estimates that there may have been as many as fifty thousand pairs in 1782, when the bird became America's national symbol. And there were doubtless more still when Columbus arrived in 1492, the year often used as a de facto baseline. So do ten thousand eagle pairs represent a miraculous resurrection, or only a meager uptick after a much longer, more devastating decline? Is America flush with eagles? Are we still hopelessly deficient? How many eagles should there be?

In 2005, a paper was published in the journal *Nature* that sought, in part, to settle this ambiguity. Its lead author was a then graduate student named Josh Donlan. If the problem of shifting baselines was starting to feel unresolvable—like nothing could be objectively measured; like we were staring down into a vertiginous, infinitely receding series of subjective baselines, each invalidated by the one just behind it—then Donlan was ready to bring everyone back onto solid ground.

Once, the paper explained, North America was teeming with spectacular prehistoric megafauna: not just Thomas Jefferson's mammoths, but shaggy eighteen-foot-long ground sloths, dire wolves, and beavers the size of bears. There were humongous elk, saber-toothed cats, wild horses, and cheetahs. There were American lions bigger than present-day African lions, and a fleet of meat-eating birds circling overhead, waiting for all these hulking carcasses to drop. Virtually all of these species vanished at the end of the Pleistocene era, about twelve thousand years ago. Why? In part, one hypothesis holds, it was because that was when humans, having arrived in North America over a land bridge from Asia, developed a new kind of stone spear tip called the Clovis point and hunted them to extinction.

Donlan and his colleagues argued that this Pleistocene extinction, and not Columbus's arrival in 1492, was North America's zero event—the moment when, ecologically speaking, everything started going wrong. If there was one scientifically defensible baseline for conservationists to agree on, this should be it. That megafauna had tremendous impacts on its ecosystems. Simply trundling around and displacing dirt would have changed the landscape in profound ways, providing habitat for other, smaller critters and insects. The absence of that megafauna has had repercussions, too. For example, it allowed the animals those larger animals ate, like deer and other ungulates, to explode in number, setting in motion a suite of other disorderly consequences.

In short, the loss of that megafauna has meant that, for the last twelve thousand years, every human generation has inherited a North America that is profoundly out of whack. So many of the ecosystems we see, study, and appreciate like architecture are, in fact, mostly ruins—a disheveled set of ripple effects, reverberating from the loss of these big and influential beasts. All of this is well understood and not especially controversial. But Donlan and his coauthors were proposing bringing that megafauna back—or at least proxies for it. Their plan called for the importation of camels, tortoises, and horses, and then eventually cheetahs, elephants, and lions from Africa. The animals would be reintroduced on private property or at new, large, fenced preserves on the Great Plains. They called the plan Pleistocene Rewilding.

People object that Pleistocene Rewilding would be playing God, Donlan told me. "But my response is, we're already playing God." We're just not doing a very good job of it. The goal of too much conservation appears to be keeping the last threads of a lost world from fraying—"managing extinction," as Donlan puts it. We are afraid to interfere with nature or steer its course too dramatically on purpose. But we are meanwhile creating a nature full of highly adaptable weeds and pests by accident. We're not just eliminating the continent's wildlife; we are oafishly changing the composition of it, replacing its condors and panthers with more dandelions and rats. At the heart of Pleistocene Rewilding was a contention that America could be so much more.

Pleistocene Rewilding became big news in the week after Donlan's proposal was published in 2005. He found himself discussing America's long-lost megafauna on morning television shows. "It was probably the biggest ecological history lesson America has ever had," Donlan told me—a revelation for ordinary people to realize that their suburban neighborhoods were once run through with camels and lions. But then Hurricane Katrina hit, and the media turned their attention elsewhere.

Donlan also got letters—hundreds of letters and e-mails—calling the idea "moronic," scolding him for wanting to "release killers in our homeland" or for proposing to pillage Africa's wildlife in order to "beautify" the Great Plains. People worried their children would be "gobbled up while taking a hike" and detailed exactly what kind of firearm they'd use to gun down trespassing elephants.

On the one hand, Pleistocene Rewilding is completely logical. If you followed history far back enough—if you peeled back all those nostalgic baselines we are burdened with, one by one—you'd arrive exactly where Donlan had. You would see the rationality of what he was proposing. But if I was learning anything, it was that rationality hardly matters when it comes to the dark comedy of people and wild animals in America. "Just a note to let you know," one man wrote to Donlan, "that those of us who actually work for a living think you are a colossal asshat."

ONE AFTERNOON, I was working at Antioch Dunes with a group of high-ranking Fish and Wildlife Service employees when an old man suddenly appeared over a small ridge, walking briskly. He wore a striped dress shirt and a fishing hat, with a pair of binoculars clipped to his belt. Blood was running down his forearm—it looked as though he'd opened up a network of scabs—but he didn't seem troubled by this. Everyone in our party turned and stared: people don't just walk around the Antioch Dunes; the refuge is usually closed to the public. Then the refuge biologist, Susan Euing, said, "That's Dr. Powell."

Jerry Powell is still a professor emeritus at Berkeley. He is a standoffish man of seventy-seven with a full head of white hair. It turns out that Powell has become one of Liam O'Brien's primary butterfly mentors. They frequently travel around California together in the summer, camping out and doing volunteer butterfly counts. When Liam gives talks about butterflies in San Francisco, Powell comes with his wife and afterward stoically points out inaccuracies. "He's got antifreeze in his blood in a curmudgeonly, really fun way," Liam told me. "I seriously think he's one of the coolest people I've ever met."

I spent a couple of afternoons in Powell's office, asking him about the Antioch Dunes. He told me that any work to recover the Lange's is decades beyond the point of diminishing returns, and even if it were possible, the agency's strategies were, in his opinion, completely misguided. For Powell, the Lange's wasn't even the point. The butterfly had been singled out arbitrarily for federal protection, but it was only a remnant of a community of insects he'd watched waste away around it. I asked him to imagine he was in charge of managing the dunes. What would he do? He couldn't answer. "So little of the habitat that made the place special is left," he said.

That afternoon in Antioch, Powell was surveying the dunes as part of an annual nationwide butterfly count—a tally of all species, not just the Lange's. Euing had loaned him a key to the refuge gate, and he'd let himself in. He walked within ten yards of our group without exactly approaching us, curious but seeming not to want to be the one who struck up the conversation.

Finally, one of the government biologists called Powell over and asked him to take a look at some buckwheat plants. The leaves were snaked with rust-colored scars, and the Fish and Wildlife biologists suspected this was because the metalmark caterpillars were crawling up the stems and feeding on the leaves. To them, it was an encouraging sign. Powell disagreed. To him, it looked like a symptom of plant disease. "I don't know what it is. It can't be doing this plant any good," he said.

The man from Fish and Wildlife stared at the plant a while longer. "Thanks for looking at that," he said.

"So . . ." Susan Euing said to Powell, trying to make conversation. That spring, Fish and Wildlife had experimented with trucking in cattle to eat the hairy vetch. Euing now told him, "The cattle are off the grazing area now. It's looking pretty good. It's working."

Powell laughed. That was his only response. He'd stuck with the Antioch Dunes for so long but seemed to have reached the same point of disillusionment with the place as his predecessors.

"Okay," Powell finally said. "Well, good luck with it." Then he handed Euing her key and left.

8.

OUR VANISHING WILDLIFE

When I got home from Antioch Dunes one afternoon that summer, there was a newspaper clipping with my name written on it, waiting for me on our kitchen table. My mother-in-law had tucked it into a package for Isla, thinking that I'd be interested.

The article described how, recently, about a hundred small diamondback terrapins had made their annual migration, crossing the runways at New York's Kennedy Airport. As happens during the summer, the turtles had delayed many flights. And the newspaper hinted at how funny it was that a few tiny turtles could halt a fleet of giant airplanes.

It *was* funny. But, then again, I'd also recently read about how Christopher Columbus's men, moored in the Caribbean after their trans-Atlantic trip, were kept awake at night by the clunking of so many sea turtle shells against the hulls of their ships. There may have been as many as 660 million green sea turtles in the Caribbean at that time—collectively, they would have weighed as much as, and maybe more than, all the buffalo on the North American plains. It was the opposite of the scene at JFK, in other words: a giant fleet of turtles was bombarding a few tiny ships.

I'd started reading historical accounts of wildlife in America—a little obsessively. Part of what I was hoping to do for Isla by showing her endangered animals in the wild was offset the environmental generational amnesia that would inevitably take hold between her and me—to help her know a world, a baseline, that preceded her. But I was also learning about the world that preceded me. And what always leapt out of those accounts was the simple fact of abundance: people's descriptions of being dwarfed and engulfed by wild animals, or even just massively inconvenienced by them. Americans are still inconvenienced by wildlife all the time, of course: the raccoon in the attic, the deer eating the backyard flowers, a recent incident of black bears rubbing against wiring in Idaho and shorting out Internet connections. But, as with the turtles at the airport, these confrontations usually highlight how inflexible our own species has become, and how much space we take up. You'd never know that the animals used to be the teeming and inflexible ones.

I read that in the eighteenth and nineteenth centuries, passenger pigeons roosted in flocks of more than a hundred million birds. Flying in, they were said to block out the sun. One man on the Ohio River mistook the "loud rushing roar, succeeded by instant darkness," for a tornado. Trees snapped under their weight, and when the birds finally moved on, the locals were left to trudge through the many inches of dung that had accumulated under them like a fetid snowfall.

But by the turn of the twentieth century, the passenger pigeon had been all but hunted to extinction. The birds' sudden absence was just as astonishing as their abundance had been. It was too much to believe. Even decades after the death of the very last passenger pigeon in

1914, a letter to *The Saturday Evening Post* speculated desperately that the pigeons were still alive and must have merely flapped off for another planet.*

I read that a man named James McNaney, while camping outside Glendive, Montana, in 1882, was suddenly confronted with "a living stream" of some hundred thousand buffalo barreling down on him from the top of a hill. The animals came running ten abreast, and still it took four hours for the herd to pass. In the early days of the railroad, trains had to stop for hours to wait for crowds of buffalo to slog past. Sometimes a stampeding column battered into the side of the train, derailing it. Describing one incident in Kansas in 1871, a man wrote: "Each individual buffalo went at it with the desperation of despair, plunging against or between locomotive and cars, just as its blind madness chanced to direct it. . . . After having trains thrown off the track twice in one week, conductors learned to have a very decided respect for the idiosyncrasies of the buffalo."

That abundance wouldn't last, either. By the early 1870s, the slaughter was well under way. The animals were killed for food, for

---

* The last passenger pigeon was named Martha. She lived at the Cincinnati Zoo. After she died, she was sent in a block of ice to the Smithsonian and stuffed. She enjoyed decades of celebrity. In 1966, Martha was sent to San Diego, on loan for a symposium about wildlife conservation; the organizers just wanted her there, as a mascot, as they plotted ways to save the earth. Fifty-two years after her death, Martha was flying again—on American Airlines. A dedicated stewardess held Martha on her lap the entire transcontinental flight. When she returned to the Smithsonian, Martha spent years as part of an exhibit of taxidermy of extinct birds, alongside a great auk, an ivory-billed woodpecker, and the world's last heath hen, named Booming Ben. But then the museum built a new Hall of Mammals, and the birds had to come down to make room. When I visited, the collections manager, James Dean, told me that he'd like Martha to be put back on exhibit somewhere, but that his PR people tell him that no one wants to see a case of extinct birds. So Martha is now kept on a Styrofoam block in a metal cabinet labeled Z-11-C, in a vast climate-controlled warehouse full of metal cabinets, reminiscent of that last scene of *Raiders of the Lost Ark.* "Poor Martha," Dean said as he pulled her out to show me.

skins, for their tongues. They were killed as an act of war against the Native Americans who relied on them. And they were killed for the sheer pleasure of shooting them, sometimes from a moving train. In 1872, after crossing much of the continent on the railroad and seeing no buffalo, President Garfield assumed that all this shooting from railcars had made the animal skittish, that the buffalo had learned to stay away from the tracks. In fact, they had just been wiped out from those areas. By the 1880s, the moping masses that once blackened the plains had been reduced to the point where the death of an individual animal was significant enough to be reported by the Associated Press. The breadth of the destruction supplied the same awe as the animals themselves had. As one newspaper put it, it was a "rate of extermination that is almost incalculable and one of which the mind can have no just conception."

It was around this time that a man named William Temple Hornaday became extremely distressed by the obliteration of the buffalo, an animal he viewed as not only a financial asset but an iconic part of America's natural glory—"our national animal," he called it. He regarded its extermination as a "national disgrace." It was also an emergency. Hornaday determined that there were fewer than three hundred wild buffalo left. And so, in 1886, he did what he reasoned to be the most helpful and logical thing: he set out for Montana to kill several dozen of them.

HORNADAY'S STORY is both an inspirational tale and a cautionary tale about the problem of shifting baselines. Somehow he withstood the tug of cynicism and despair that I'd seen drag down each generation of Bay Area lepidopterists. Rather than eulogizing the wildlife vanishing around him, Hornaday scrambled to defend it, and even replace it. For a time, at least. Maybe it was because he fought back that disillusionment for so long that when it finally did get a purchase on him, it took him down hard.

Hornaday was a taxidermist—a gifted one. He was born in 1854; beginning in his late teens, he made many perilous trips around the world to hunt exotic animals to stuff. (He claimed that during these trips he survived a jaguar attack, wrestled a crocodile, procured an orangutan named Little Man to give Andrew Carnegie as a gift, and sailed past a manta ray the size of a small volcanic island.) By the time he staged his buffalo-hunting expedition to Montana, he had ascended to a job at the Smithsonian. He was like America's taxidermist laureate.

Hornaday needed to kill buffalo so that he could preserve them. He was horrified by the lack of buffalo specimens at the nation's premier museum. And though he acknowledged that the idea of shooting any of the remaining wild ones was "exceedingly unpleasant," he felt it had to be done, so that people in the future could still see and know them.

He and his party took twenty-five buffalo on a trip he billed as "The Last Buffalo Hunt." Back in Washington, Hornaday arranged the choicest animals into a scene around an alkali watering hole. He considered these the wildest and fittest buffalo, since they had survived the great extermination and outlasted all others. "For years," he wrote of one bull, "the never-ceasing race for life had utterly prevented the secretion of useless and cumbersome fat." While preparing the animals, Hornaday had discovered at least one older bullet lodged in nearly every adult; the only buffalo left on the range, it seemed, were those that had survived previous assaults.

By then, Hornaday had become fixated on starting a zoo. Zoos were relatively new ideas in America—the nation's first had opened in Philadelphia in 1874. Hornaday saw their potential as, essentially, galleries of living, moving taxidermy—safe houses where at least a few animals could be preserved and also bred, so that their kind would never technically die out. To that end, he was instrumental in establishing a Department of Living Animals at the Smithsonian, which later became the National Zoo. And in 1896, he was tapped by the New York Zoological Society to direct its new zoo, what we know as the Bronx Zoo. He stayed in that job for the next thirty years. Hornaday insisted that the zoo be open to the public for free. He envisioned it as a place where ordinary Americans could experience wildlife in perpetuity, no matter how many more extinctions occurred outside its gates.

Reconnecting the American public to its wildlife would be the major theme of Hornaday's life. It frightened him that, as the country urbanized, younger generations were losing their firsthand knowledge of animals. He was one of the nature fakers' brashest critics, condemning them for filling that void with fairy tales. And he similarly attacked scientific education's growing emphasis on cellular biology at the expense of getting outdoors and learning the natural histories of wild animals through observation. For Hornaday, there was a deep tie between American wildlife and American greatness. His biographer Gregory Dehler writes that the country's new fascination with microscopes struck Hornaday as "downright unpatriotic and un-American." It missed the aesthetic dimension of wildlife. Animals were big and awesome; cells were slow and boring. Animals were accessible to anyone; cells were for eggheads. In theory, Hornaday was an ardent populist, though he often had trouble putting up with the actual populace. Scandalized by people freely dropping their trash around the Bronx Zoo after lunch, for example, he complained in a letter to the *New York Times* that America was becoming a society of litterers—then removed most of the benches, so that there was nowhere to eat lunch anymore.

Ultimately, it may be impossible to fix into a consistent shape all the eccentricities and outlandish opinions of Hornaday's life. Little about him meshes with our modern ideas about wildlife conservation, ecology, or political correctness. He blamed the decline of the buffalo partly on the animal's "own unparalleled stupidity." The anthropomorphism of animals sickened him, except when it didn't. (He once wistfully remembered an orangutan "friend" named Dohong whom he described as "an engineer, an investigator and a philosopher.") And he used his observations at the zoo to work up a chart titled "Estimates of the Comparative Intelligence and Ability of Certain Conspicuous Wild Animals, Based upon Known Performances, or the Absence of Them," in which he rated twenty different species on a scale of zero to a hundred in categories like "Perceptive Faculties," "Nervous Energy," and "Use of the Voice." (The beaver, for example, scores one hundred in "Original Thought.") He was also a racist. In 1906, he put a pygmy man from the Congo named Ota Benga on display in a cage at the zoo.

Still, it's not hard to see a peculiar heroism in his life's work. So many stories I stumbled on while learning about the Lange's metalmark followed the same arc: people losing hope as, through their lifetimes, they watched the animals disappear and the baselines shift. Meanwhile, baselines for success were getting defined downward, too, with conservationists laboring to safeguard and cherish what those who came before would have regarded as scraps.

But Hornaday's thinking about wildlife conservation seemed to only grow progressively more optimistic and ambitious as time wore on. It occurred to him that *more* of the world he inherited could be saved, not less. He'd started out wanting to save buffalo for future generations by stuffing them, then evolved to build a reliquary of living animals at a city zoo. Finally, he devoted himself to something even more audacious: putting actual buffalo back out there, on the plains.

In 1905, nineteen years after his buffalo hunt in Montana, Hornaday founded the American Bison Society with his friend President Theodore Roosevelt. The group rounded up surviving buffalo and brought them to Hornaday's zoo to be bred in captivity. By 1907, the society was working with the federal government to ship batches of animals to large, fenced preserves on the plains. It was wildlife conservation's first captive breeding and reintroduction program, a cornerstone of many endangered species recoveries today.

The public rallied behind the buffalo project. This was, after all, only five years after Roosevelt's hunt in Mississippi had birthed the teddy bear, when America was starting to reckon with having torn apart a continent's worth of wild animals—and to romanticize, and not only fear, whatever was left. Now William Temple Hornaday, nattily dressed in suit and top hat, was boldly crating up bison in New York City so that they could once again fill the Great Plains, shipping them out on the same railroads that had helped obliterate the animals. It felt like a reconciliation, as though an injustice was starting to be incrementally righted. The *New York Times* wrote that Hornaday "deserves the gratitude of the Nation."

From there, Hornaday's ambitions only expanded further. Backed by a discretionary fund established by wealthy friends, he lobbied Congress for landmark laws to protect fur seals and migratory birds. He also lobbied the public directly. In 1913, he tried to snap the American people into action with his most famous book, an encyclopedic screed titled *Our Vanishing Wildlife: Its Extermination and Preservation.*

It had been more than a century since Thomas Jefferson began *Notes on the State of Virginia,* writing to far-flung acquaintances, collating their observations about the bigness of American fauna—

evidence with which he could combat Buffon's Theory of American Degeneracy. Now Hornaday did almost the opposite. He asked game wardens and wildlife experts across the continent for firsthand reports of how badly that wildlife had been whittled down. "The sandhill crane has been killed by sportsmen," a man in Minnesota wrote. "I have not seen one in three years." Another reported that white-tailed deer were entirely gone from Delaware. "Antelope, mountain sheep and grizzly bears are *going*, fast!" one operative in Wyoming wrote Hornaday. Even the rabbits were on the ropes.

Our Vanishing Wildlife became a bestseller. It was an antagonistic and panicky book; at times, the pages seem to be riveted together with exclamation points. Hornaday covered the decimation of all species in the same frenzied tenor, from the iconic grizzly and the whooping crane, down to the pheasant and the squirrel. ("A live squirrel in a tree is poetry in motion; but on the table a squirrel is a rodent that tastes as a rat smells," he wrote. "We ask every American to lend a hand to save Silver-Tail.")

But underlying his alarmism was a radical idea. Previously, concern about disappearing wildlife focused on game species and the desire to keep them perpetually in supply for hunters. But Hornaday cast hunters as his villains. Although he was not antihunting, the wholesale destruction that was now taking place, enabled by more powerful firearms and the better access to wildernesses granted by cars, struck him as gratuitous and pointless. The damage being done would be irreversible. Well more than a century after the American Incognitum debates, Hornaday still felt it necessary to stress: "It is time for all men to be told in the plainest terms that there never has existed, anywhere in historic times, a volume of wild life so great that civilized man could not quickly exterminate it by his methods of destruction."

Hornaday was arguing that the value of wild animals in America wasn't just utilitarian. He spoke to the growing segment of the public who did not hunt and likely lived in cities and who, he insisted, shared just as equally in the ownership of those animals. He was democratizing wild animals, exposing the interests of hunters as narrow. There were other relationships to enjoy with an elk or a cougar besides shooting it—aesthetic or even imaginative relationships that respected their "value as living neighbors to man."

Throughout *Our Vanishing Wildlife,* Hornaday seems to be trying out new ways of talking about extinction that might instill an emotional sense that even a country growing so detached from nature had something to lose if more of that nature disappeared. ("The birds and mammals now are literally dying for *your* help," he wrote. "Can you not hear the call of the wild remnant?") He was pioneering a new, more emotional style of conservation. And though this approach laid the groundwork for so many modern campaigns—prefiguring, for example, the empowering but also gently accusatory stump speeches I heard Robert Buchanan give about our duty to polar bears while on Buggy One—it only estranged Hornaday from other wildlife advocates of his time. He was criticized as "the Bolshevik element of game conservation" and as effeminate or senile. One historian describes him as being purposefully edited out of the story of American environmentalism.

Hornaday fought on for two more decades, however. By 1930, his American Bison Society had stocked six different preserves around the plains with herds of pure-blood buffalo. He deemed the animal's future "absolutely secure." But he'd turned pessimistic about so much else. During World War I, Hornaday had watched humanity take the high-tech artillery it was using on wild animals and turn it on itself, and he'd never really recovered from that terror. The war spoiled his faith in people's decency and wisdom. He worried about overpopulation. He worried about the Chinese. Mainly, he was dismayed that a second world war seemed inevitable.

As he grew older, it became almost impossible for Hornaday to control his dread. Introducing a mostly whimsical book about animal intelligence, he recommended America read it now, "before the bravest and the best of the wild creatures of the earth go down and out under the merciless and inexorable steam roller that we call Civilization." And at the end of his life, he wrote in an unpublished memoir: "Thirty years ago, I was a sincere optimist on the impulses and good-faith of humanity, and the moral fiber and intelligence of civilized man. Today, I think that speaking generally, Civilized Man is an unmitigated ass."

I was beginning to notice a corollary to the concept of shifting baselines syndrome, one that doesn't seem to be discussed in the literature. It's not just that we start our lives unaware of the damage that preceded us, but that we end them burdened with having witnessed so much damage done. The clean slate we inherit gets mucked up all over again. Maybe there's a tipping point in every life when that muck is finally too much.

Hornaday died in March 1937. The local Boy Scout troop's color guard surrounded his coffin, and the buglers played "Home on the Range."

AMERICA, HOWEVER, has a way of assimilating revolutionary ideas while forgetting the revolutionaries who originated them. Beliefs that Hornaday had been teasing out in *Our Vanishing Wildlife*—that there is some intrinsic and ineffable value to wildlife; that a moral obligation to care for it has been placed on us—would be at the heart of the Endangered Species Act, too. And even by the time the act passed in 1973, less than forty years after Hornaday's death, those beliefs had become mainstream enough that affirming them struck virtually everyone in Congress as a wholly uncontroversial and feel-good bit of politics. Signing the Endangered Species Act that December, President Richard Nixon issued a statement that, with its exaggerated patriotism and determination, could have been lifted out of Hornaday's book: "Nothing is more priceless or more worthy of preservation than the rich array of animal life with which our country has been blessed," Nixon said. "It is a many-faceted treasure, of value to scholars, scientists, and nature lovers alike, and it forms a vital part of the heritage we all share as Americans."

That heritage, I was now learning, is mostly made up of bugs. Well more than half the species on Earth are insects—an estimated five to thirty million insect species, though we've only discovered and named about a million of them. Many of those insects have critical impacts on their ecosystems. They are the Pleistocene megafauna writ small. They chew up and decompose the dead to keep things clean and keep energy circulating through its natural cycle. They riddle the soil with holes to aerate it. They spread seeds. They pollinate a third of the foods Americans eat. They are useful, in other words. But the large majority are also characterless and ugly—not quick to draw our sympathy. And even though the Endangered Species Act that Nixon signed did quietly entitle imperiled insects to the same protection as sexier critters like bald eagles and blue whales, the agency then overseeing the listing process, the Office of Endangered Species, was shy about listing any. Wildlife conservation was still a new and controversial idea; insect conservation was barely an idea at all. Eventually, the lone entomologist inside the Office of Endangered Species seized on butterflies as a safe way to test the political waters. By 1976, only two insects had been protected as endangered species—both of them butterflies. Then, on June 1, 1976, six butterflies in California were listed simultaneously. The Lange's metalmark was one of them.

At first, officially protecting the Lange's metalmark proved to be the very worst thing for it. The Antioch Dunes were privately owned, and the landowners, assuming that the government would step in to buy the dunes or restrict their use, started selling off the remaining sand there to miners, trying to wring as much value from their property while they could. Truckloads vanished every day. "Sand was being hauled away literally as we were catching butterflies out there," a lepidopterist named Richard Arnold remembers.

Arnold was a fixture at the dunes in those days, studying the Lange's as part of his dissertation research at Berkeley. Very little was known about the butterfly, and Arnold was collecting the basic scientific information that the Fish and Wildlife Service would need to recover the species. He spent nine summers on the property, tracking the butterflies' behavior. Over time, he meticulously diagrammed the flight patterns of more than eight hundred individual butterflies by drawing different identifying patterns on their wings with a Sharpie. Often he'd arrive in the morning to find that the sand of one of his study sites had been mined out from under him.

In the summer of 1979, a landscaper hired by the utility company to maintain the land it owned under the power line towers accidentally rototilled through four or five hundred buckwheat plants—one of two last major butterfly colonies. Ironically, until then the terrain around the towers had remained some of the best-preserved butterfly habitat precisely because it had been built on. For decades, the sand miners had been forced to work around the towers. Gradually, they had dug a wide, deep pit between them, so that by 1979 the towers stood on what are still the most conspicuously dunelike formations left: two spectacularly steep hills, reaching some seventy-five feet in the air. Today the pit opened between those hills has been invaded by shrubs and grasses. Oak trees have put down roots and grown twenty or thirty feet tall; walking around, you feel like you're in the woods.

THE U.S. FISH AND WILDLIFE SERVICE bought the Antioch Dunes in March 1980 for $2.2 million. It had taken more than a year to negotiate the purchase of the land. There were competing offers from condominium developers; a proposal for a "Sand Dunes Waterfront Park," with a marina, fishing pier, campground, swimming lagoon, and science center; and objections from the city government, which preferred to see a revenue-generating riverfront showpiece built on some of its last accessible waterfront. While everyone haggled, the sand miners kept working, until, by the time a deal was cut, there was little sand left to sell. That summer, Richard Arnold estimated that the Lange's population had collapsed to eight hundred butterflies, down to half of its size two years earlier. But now, with the establishment of a national wildlife refuge at Antioch Dunes, the government seemed to have beaten back humanity and protected the habitat just in time. The abuse the butterfly had weathered seemed to be over. It wasn't.

There is an assumption, writes the environmental legal scholar Holly Doremus, that "what nature needs most is for people to leave it alone"—that a landscape will "automatically produce the preferred human outcome, a perfect Garden of Eden, if it is simply walled off from human influence." But nature doesn't know what outcome we want, and it doesn't care. Instead, it perpetually absorbs what we do or don't do to it, and disinterestedly spits out the effects of those causes. Nature is not a photograph that will always look good if we keep our fingerprints off it. It's a calculator, adding up numbers we don't always realize we're pressing and confronting us with the sum.

Doremus describes the example of the Hutcheson Memorial Forest, a sixty-five-acre tract of old-growth oak and hickory trees in central New Jersey, which was set aside as a preservation area in the 1950s.

But foresters eventually noticed that maple trees were overtaking the forest's signature oaks and hickories. Analysis of tree rings suggested that Native Americans had regularly set fire to the woods, until about 1700, probably to flush out animals while hunting or to maintain travel routes. These fires would have killed off any new maple tree growth, allowing the well-established oaks and hickories to continue to dominate. In 1954, *Life* magazine did an extravagant spread about the Hutcheson Forest, with a series of Disneyesque paintings showing the mourning doves, red fox cubs, downy woodpeckers, toads, raccoons, fawns, rabbits, song sparrows, and beavers that lived in the woods. The area was meant to be a living diorama of a natural world that we'd otherwise lost, now that "man has colonized the planet from its white polar regions to the hot midriff of the equator." But when we left the forest alone, the nature that took its course there didn't look like the nature in the paintings.

At Antioch, nearly a century of sand mining had shredded an equally complicated ecology. By the time Fish and Wildlife stepped in, gusts of wind still lifted up what sand remained, but with no sloping dunes there to catch it, it couldn't pile up. It merely scattered into the river or across the street, dispersing out of the system forever. Leaving the dunes alone at that point would only have allowed them to keep unraveling. The original order—the moving mosaic of sand, plants, and butterflies—had to be either set back in motion or forever simulated. We had to do the opposite of leave the ecosystem alone. We had to disturb it. As one lepidopterist who has worked on the Lange's metalmark recovery told me, "We can't just throw up a fence and think everything's going to go back to how it used to be." Still, that's almost exactly what happened after the government bought Antioch Dunes.

The dunes were the first national wildlife refuge set aside for only insects and plants. The Fish and Wildlife Service had less experience with bugs and shrubs, and, operating outside its comfort zone, the agency worried that interfering to improve the landscape for butterflies could, in some way, wind up damaging it for the two endangered plants. There were a few attempts to plant new expanses of buckwheat, but Fish and Wildlife was largely paralyzed.

Meanwhile, all the controversy prior to the government's purchase of the property had raised the profile of Antioch Dunes. Before, locals hadn't thought much about the land out by the gypsum plant. Now guys on dirt bikes vaulted off the remaining hills. More fishermen, vagrants, and campers turned up. A few campfires got out of control. In 1984, a butterfly poacher was apprehended there after netting at least one Lange's.

For anyone rooting for the Lange's metalmark, the first official years of the Antioch Dunes National Wildlife Refuge were frustrating—maybe a little disillusioning. Then, in October 1985, there was a catastrophe.

THE WHALE WAS forty feet long and estimated to weigh forty tons. It was a humpback and diverged from its migration route along the Pacific coast, from Alaska to Mexico and appeared to have entered San Francisco Bay sometime on October 11.

The whale kept swimming. It headed northeast, inland, with great purposefulness, as if it knew where it was going and was running late. It swam into the Carquinez Strait, then the mouth of the San Joaquin River. It kept going. A week later, it had traveled under a series of bridges and more than sixty miles upriver, into the Sacramento River, and halfway to the state capital. Then it let up and lingered outside the small town of Rio Vista.

Biologists were convinced that if the whale stayed in the river much longer it would die. There was nothing for a whale to eat in a river, and as the water got less salty upstream, the animal would become less buoyant and expend more energy to stay afloat. Its skin already looked pallid and its breathing seemed irregular. But, truthfully, not enough was known about humpback whales in 1985 for anyone to say for sure, or to explain what the hell the animal was doing. "Perhaps he's insane," one bearded scientist told *Nightline*. By this time, the whale was a national story. A scattershot armada of government officials, biologists, and several hundred local volunteer boat owners coasted alongside it every day, trying to turn it around. Everyone called the whale Humphrey the Humpback.

"It was like Woodstock," one *San Francisco Chronicle* reporter later recalled. Thousands of people lined the riverbanks. Entrepreneurs sold food and T-shirts. When Humphrey spent a few days in a narrow, shallow irrigation canal called Shag Slough, a shop in Rio Vista started selling mimeographed maps to the slough for a dollar, and crowds drove out to see the unthinkable: a living whale, up close, with a cow grazing behind it.

By the end of the second week, hundreds of unsolicited suggestions were flooding in from kids in science classes around the country and ordinary can-do Americans. (These included building a sexy papier-mâché female whale to attract Humphrey downstream.) At one point, a boatload of new-age parapsychologists coasted in behind Humphrey, seeking to generate a benevolent, telepathic force field to nudge him homeward. A money-market fund was set up in Humphrey's name to cache donations. There were rumblings about Wayne Newton staging a benefit concert. One night, a local construction company cleared underwater debris out of Humphrey's path. "It was a magical time," Diana Reiss, a biologist who helped direct the rescue, told me. "You'd say, 'We need *this* to save Humphrey,' and you'd get it."

When people remember Humphrey today, they tend to preface their stories by describing how sullen the mood in America was in

October 1985. TWA Flight 847 had been hijacked that June—a United States Navy man was tortured and dumped dead on the tarmac—and, only four days before Humphrey entered San Francisco Bay, four terrorists from the Palestinian Liberation Front hijacked the ocean liner *Achille Lauro,* throwing an American retiree in a wheelchair overboard.

People made the point explicitly at the time: America needed this whale, needed something to rally around and save. It was easy to identify with the lost animal: helpless, in an unfamiliar and dangerous world, unable to find his way out. He'd left the ocean—maybe the last uncorrupted-seeming and mysterious region of the earth—and bored directly into this uglier human realm of ports, oil refineries, and heavy manufacturing plants. His very presence in our world seemed troubling and wrong. He deserved better than to be around us. So everyone, united in kindness, was trying to get Humphrey home.*

For the most part, the rescuers had been experimenting with ways to frighten Humphrey downstream. Eventually, one Sunday morning more than three weeks after Humphrey had left the Pacific, they decided to try to *attract* Humphrey back to the ocean instead. They outfitted a fishing boat called the *Bootlegger* with underwater speak-

---

* In retrospect, it's amazing how the idiosyncratic motivations of so many people aligned behind a single whale. For example, the lead boat for most of the rescue effort was a small fishing vessel called the *Sportfish I,* captained by a local fishing guide named Jack Findleton. As a teenager, Findleton had seen twelve months of combat in Vietnam. He was one of only 27 of the 144 men in his unit who came home alive, and he came home scarred. "This is my way of making up for what I did then," he told one newspaper, an atonement for some apparently very terrible acts, which he would not go into. Trying to save Humphrey, Findleton said, "has shown me sensitive feelings that were buried for years." Another man involved in the rescue mission, Bernie Krause, told me that, years later, a woman in Ohio explained to him that she and her family and everyone in their church had been praying for Humphrey. They believed the animal was Jesus Christ, returned in the form of a whale, swimming up the river to take a look around and judge us.

ers and broadcast recordings of other humpback whales feeding. It worked. That day, the humpback would wind up following the *Bootlegger* for eleven hours, with an armada of thirty other boats falling in behind them to keep him from turning back. Finally, late the following afternoon, just in time to make the evening news, Humphrey was led under the Golden Gate Bridge and back into the Pacific.

It had been apparent that the new strategy would work almost immediately when the armada of rescue workers convened upriver that first morning. ("It looks like the Normandy invasion," one woman said as the operation got under way—by now it included helicopters, an eighty-two-foot-long Coast Guard ship, military riverboats, and the usual long tail of houseboats, fishing boats, and dinghies.) As soon as the whale call recording was switched on, Humphrey surfaced out of nowhere and locked onto the *Bootlegger*. Someone on the riverbank with a boom box blasted "Born to Be Wild" as the whale, tight on the *Bootlegger*'s tail, spouted a plume of water vapor and snot into the air and—finally moving toward the ocean now—rolled under the Antioch Bridge and past the Antioch Dunes.

All of this had been unfolding in plain sight of the Antioch Dunes National Wildlife Refuge. In fact, Humphrey had been dawdling in the area for several days, and all weekend people had flocked to the dunes to see the whale swim aimlessly back and forth, swatting the water with his fins, shadowed by his beleaguered entourage. The refuge was some of the only undeveloped, open riverbank left where people could stand and watch. Five thousand people were counted on one half of the property in a single afternoon—a mob that parked and walked where it pleased, ignoring signs and the small detail of refuge employees who had rushed to the dunes to try to instill order. Now, as soon as Humphrey and the *Bootlegger* got going, motoring past the Antioch Dunes and heading home, the crowd rushed into their cars and peeled off for the next viewing opportunity in a neighboring town. "The place cleared out like the plague had hit," one refuge employee reported.

The dunes had been trampled. Endangered plants were destroyed, new trails had been stamped into the sand, and existing ones were smeared open and destabilized to the point where, a week later, after heavy rains, an entire section of hillside collapsed into the river. With it went many buckwheat plants and whatever Lange's metalmark eggs had been laid on them that summer. An internal Fish and Wildlife Service memo noted, "It is ironic that one endangered species"—the humpback—"could have such an impact on three other endangered species."

It was a tragedy of charisma. Because of their affection for a single celebrity whale, hordes of people had jeopardized an entire species of anonymous butterfly. They weren't an angry mob; they were a loving mob. But they loved only certain things.

FISH AND WILDLIFE'S response to Humphrey was swift and, ultimately, maybe more damaging than the crowd itself. The agency put a fence around Antioch Dunes, locking out the dirt bikers, fishermen, campers, and butterfly poachers, as well as about a thousand hikers and picnickers who visited the dunes legally during the year. If you believe that nature needs us to leave it alone, this sounds like an unequivocally good thing; but in retrospect, that traffic was actually providing the last vestiges of disturbance in the disturbance ecosystem: wearing down dunes and shifting around sand to keep the niches for native plants like the buckwheat open. Maybe they were providing too much disturbance, or disturbing too recklessly. But entomologists I spoke to argue that even the stampedes to see Humphrey may have been only simulating the kind of dramatic dune collapse that would have happened there occasionally for the last one hundred and forty thousand years. Now, with nothing to disturb the dunes, the landscape got almost entirely socked in with a circus of weeds. As a consequence, by 1987, peak count for Lange's metalmark slipped to 248.

Scrambling to secure Antioch Dunes after the Humphrey debacle, a new biologist at the Fish and Wildlife Service also ordered Richard Arnold to terminate his research on the butterfly. (According to Arnold, she felt his study—his handling butterflies and scribbling on them with Sharpies—was only further damaging the species, even though Arnold had data proving his techniques were safe.) Arnold had started studying the Lange's while at Berkeley, under the mentorship of Jerry Powell. In fact, Arnold had helped Powell with his years-long study trying to sketch out the ecosystem's decline. By now, Powell was well on his way to feeling that the butterfly's case was hopeless. But Arnold, as part of a younger generation, was still optimistic; he hadn't written off the dunes. He had reams of data, not just about the butterfly, but also about the refuge's two federally protected plants. Powell had even more data. Arnold felt they were working toward offering Fish and Wildlife everything it needed to make a blueprint for restoring the dunes and balancing the needs of Antioch's three endangered species. At the very least, Arnold might offer the government a meticulous account of one butterfly species' meandering road to extinction, which would bear lessons for saving others. "I think they lost a golden opportunity," he says.

When asked to leave, though, Arnold wasn't motivated to put up a fight. In truth, he was already looking for an exit strategy, searching in vain for a graduate student to take over his research and molder in the Antioch heat without pay, for only the love of the butterfly. He was discouraged by what he saw happening to the habitat. But the real problem was, Arnold had a rash. After years of prolonged exposure, he'd developed an allergy to the gypsum dust that, in those days, poured over the dunes from the wallboard factory next door. "It was a nasty situation," he says. It required frequent steroid shots. "I wanted to continue the work, but I was suffering so badly." In the last twenty-five years, Arnold has consulted on butterfly conservation projects up and down the state of California, but he's been back to the Antioch Dunes only a couple of times. "I can't risk going out there."

After that, in the nineties, management of the Antioch Dunes proceeded in sometimes constructive fits and starts. At one point, a huge garden of buckwheat was planted. Truckloads of new sand were dumped on the property and sculpted into artificial dunes. But the landscape had a way of shrugging off any improvements and jostling back into disarray. The butterfly numbers bobbled up and down.

We don't think of evolution as being steered by chance. But just as the story of the dinosaurs' extinction starts with a freakish meteor crash, the recent natural history of the Lange's metalmark butterfly was jagged off course by a humpback whale and a man's rash. It wasn't until recently—when peak count plunged into the double digits—that Fish and Wildlife hired an ecologist named Jana Johnson to help undo those decades of damage and neglect.

9.

WITHOUT CHANGE, THERE WOULD BE NO BUTTERFLIES

Every August since 2007, Jana Johnson has captured a handful of female butterflies at Antioch Dunes, transferred them into plastic containers, and secured them in the backseat of her SUV like small children. She then drives the Lange's 350 miles south, to Moorpark College, the two-year community college where she teaches in Ventura County, outside Los Angeles.

The idea is to capture female butterflies that appear to have already mated in the wild. (They tend to have enlarged abdomens and look a little banged up.) Jana calls them "foundresses." Having mated once, they will continue to lay eggs in captivity until they die. Jana and her students work to harvest as many eggs as they can, then rear them through the winter into butterflies. As in William Temple Hornaday's buffalo reintroduction, they are laboring to pump out large numbers of an endangered species in a controlled environment, then releasing them into the wild to bolster the faltering population. Her first summer on the job, Jana managed to keep one foundress alive in Moorpark for twenty-eight days. The butterfly laid more than two hundred eggs in that time. It died while laying an egg, in fact. "She was a good lady," Jana told me. "She went down swinging."

Jana calls her outfit the Butterfly Project. It occupies a narrow, rectangular yard on a hilltop of the Moorpark College campus. The Butterfly Project is wedged inside America's Teaching Zoo, a fully functioning zoo, housing about 125 animals, that serves as a noisy laboratory for students in the college's exotic animal training program. It's here that young men and women who dream of directing a troupe of performing capuchin monkeys at Universal Studios, or wrangling the live lions at the MGM Grand in Las Vegas, come to learn their trade. (The degree program is called Exotic Animal Training and Management, which all the students call EATM for short. They pronounce it "Eat 'em," and I never once heard someone acknowledge the cruel, karmic possibilities of giving that nickname to a program for lion-tamers-in-training.) It's not uncommon while at the Butterfly Project to see a potbellied pig, or Spirit the mountain lion, or a humpy white Indian cow called a zebu pace by, out for a walk with its new student trainer, establishing a bond. A few of the younger animals at America's Teaching Zoo are on loan from Hollywood movie studios, sent to Moorpark to be broken in. Others are retired animal actors. One morning, I met a turkey vulture named Puppy who had a cameo in *Airplane!*

It was August, and Jana had delivered that year's foundresses from Antioch Dunes only a few days before. With peak counts so low, she was permitted to catch only four butterflies. Meanwhile, the eggs laid in captivity the previous year were tearing free from their cocoons. There were now twenty-six adult butterflies in-house, and a twenty-seventh would wriggle out just before lunch. Each one had to be fed two or three times a day, by hand, with a Q-tip soaked in honeywater—an aggravating job that wound up taking all morning.

In the afternoon, the foundresses were transferred onto potted buckwheat plants so they could keep laying eggs. (Students assemble homemade enclosures for each butterfly by shoving the buckwheat through the bottom of an upside-down quart-sized clear plastic deli container from a Smart & Final grocery store, ventilating it, and sealing it off from pests with duct tape and toilet paper.) The other butterflies, meanwhile, spent their afternoon split into groups of four or five. The hope is that these adults will pair off and breed, and students sign up for one-hour "mating watch" shifts to keep them under observation. It's monotonous work; the kids are allowed to bring a book, but Jana requires them to look up and check for butterfly sex every time they turn a page.

The students also have to move the containers frequently from place to place, in and out of shade, to keep the butterflies inside from getting too hot or too cold, and to try to catch a certain mysterious quality of dappled sunlight that appears to put the insects in the mood. And so, all afternoon, Moorpark students stare at butterflies, periodically stand up, pick up their designated deli container, walk a few paces, and sit down in another ratty collapsible vinyl chair, where they resume staring blankly into the container as though it were a campfire—all while the car alarm–like shrieks of the zoo's primates occasionally go off in the background. "We call it walking the butterflies," I heard Jana explain while training a new student named Paul, preparing him for the tedium of mating watch. Paul was unfazed. "I inspect a thousand fruit cups at work every day," he told her. "I totally know what you mean."

Jana is forty-one, with long blond hair and a playful presence that's more summer camp counselor than biology professor. She calls the butterfly larvae "chunky dudes," and high-fives students who report in good news. She grew up in Austin, the daughter of a Lutheran pastor, and there's a winning guilelessness about her that sometimes oozes into sentimentality about her work. (I once heard her tell a couple of Fish and Wildlife bureaucrats out of the blue about a lovely quote she'd found online: "Without change, there would be no butterflies." "I like that," Jana said, smiling.) I was both impressed and surprised that she could feel such romantic conviction, given the condition of Antioch Dunes and of her makeshift butterfly operation at Moorpark, all of which was only making me ambivalent. One afternoon, over frozen yogurt at a strip mall down the hill from the zoo, I told Jana as much. The Lange's was forcing me to ask a question I'd mostly been trying to duck: why save any species? Compared with the polar bear, the butterfly had no public profile and middling charisma. It was hard to know who would miss it.

"Intrinsically, I just know I'm doing the right thing," Jana told me. Still, she's learned that you can justify saving butterflies intellectually to different people in different ways, and she started to run through some answers. She talked about "Spaceship Earth," the idea that the planet is a spacecraft we are all living on, and that each species we destroy represents a rivet falling out of its hull. Eventually, so many rivets can be lost that the craft crashes. Wildlife does work to keep the planet functioning. And protecting individual species has proved to be a way to protect, or even repair, entire ecosystems. (Conserving top predators like wolves, for example, can initiate powerful changes all the way down a food chain, snapping the entire landscape back into balance.) This argument for biodiversity is compelling enough that an entire field has developed to itemize the work that species do and put a monetary value on their "ecosystem services." (One recent study pegged the net worth of bats, as insect-eating helpers to America's agricultural industry, at a minimum of $3 billion.) And yet the Lange's is clearly one of the many species that the world would get along fine without. Even if it were an invaluable constituent of its ecosystem, that ecosystem has changed so severely that putting the butterfly back couldn't correct it.

Jana tried another tack. She is a single mother—I'd noticed artwork by her two young sons hanging in the Butterfly Project greenhouse—and she told me about a recent trip she'd taken with her boys, driving cross-country from California to Boston. She'd gone out of her way to eat and shop at locally owned businesses, she said, to show her boys that different parts of America were, in fact, different and unfamiliar from their experience; that the United States was not one contiguous Los Angeles. There are organisms that are doing exceptionally well as humans take over the planet, Jana said: rats, pigeons, starlings, roaches, kudzu, jellyfish. Some of them are pretty cool. But they're like ecological Applebee's and Walmart, she said, spreading through nature and homogenizing it, while putting the more fragile mom-and-pops out of business.

One of America's first butterfly conservationists, Robert Michael Pyle, has written about what he calls the "extinction of experience"—the idea that, while it may be important to protect species that are on the verge of disappearing forever globally, it's also crucial to maintain the biological diversity that we find around us, locally. As Pyle's hometown in Colorado grew from rural to suburban, many of the butterfly species he used to catch as a boy disappeared. None went extinct—they all still flew elsewhere in the United States, and sometimes abundantly—but what did go extinct is the experience of seeing them in that town and of living among them. The animals we encounter in childhood serve as our "windows on the world," Pyle writes, and "a face-to-face encounter with a banana slug means much more than a Komodo dragon seen on television." In that sense, if a species vanishes from within our immediate reach, it may as well be extinct worldwide. And if we don't personally experience biodiversity, we won't expect biodiversity to exist anywhere, or be sad to learn it is disappearing: if our baseline starts out dishearteningly low, we'll hardly be alarmed when it shifts further.

"The point," Jana told me, "is to keep some uniqueness in the world." Not just for her sons, but for their sons, too.

JANA WAS COMFORTABLE working in places like Antioch Dunes, able to see their promise. Before the Lange's, she'd devoted herself to another beleaguered butterfly, condemned to an even grimmer landscape.

In 1994, a UCLA lepidopterist named Rudi Mattoni and two colleagues happened upon a remnant population of a gleaming blue butterfly called the Palos Verdes blue. The blue had once flown widely in Southern California, on coastal scrub. But it was believed extinct since a municipal baseball diamond was built over its last known habitat in 1983, supposedly by accident. Now Mattoni had found a population on a 331-acre Southern California naval facility known as Defense Fuel Support Point, San Pedro, built to store and distribute fuel during World War II. It's a field of underground fuel tanks, next to an oil refinery that periodically ejects glowing gas flares several dozen feet into the air. Mattoni's rediscovery of the butterfly occurred right around Easter. The national media portrayed it as a hopeful allegory of resurrection in an especially hopeless-seeming place.

As strange as it sounds, military bases are actually proving to be strongholds of biodiversity. Land owned by the Department of Defense now has more endangered species on it per acre than land owned by the Department of the Interior, the arm of the government that is actually responsible for setting up refuges to conserve those species. And the military prides itself on the number of endangered species that have survived on these large hunks of habitat, accidentally kept intact as bases and maneuver sites, and is working actively to sustain them. These animals include many rare and endangered birds—Kirtland's warblers, red-cockaded woodpeckers, brown-headed nuthatches, sulfur-bellied flycatchers—and butterflies, including the Taylor's checkerspot, largely concentrated in Washington State, on a grassland abutting a live-fire artillery range. In 2008, two former colleagues of Mattoni's, Travis Longcore and Catherine Rich, wrote an essay that celebrated these run-down habitats, as well as "derelict and degraded" urban fragments like Antioch Dunes. Though they may look nothing like our romantic images of nature, they are important sanctuaries for imperiled insects, and especially butterflies. The essay, titled "Invertebrate Conservation at the Gates of Hell," argued that we should get over our queasiness about such places, and undo the strain of biblical thinking that sees everything but big, unbroken wildernesses as fallen from grace. Beginning to care about the butterflies sticking it out in these unspectacular niches can help us confront the world as it actually is, not how we'd still like to imagine it—to see nature in its modern context without bitterness and with a sense that lots can still be done.

After rediscovering the Palos Verdes blue at the fuel depot, Rudi Mattoni was put in charge of a captive breeding operation for the butterfly. In 2003, he hired Jana as his assistant. She had moved to Los Angeles from Austin several years earlier as a newlywed. Her husband was going to try to make it as an actor, and she was starting graduate school in ecology at UCLA. But before long, her marriage ended, devolving into what would become a contentious and drawn-out divorce. Pregnant with her second child, she was desperate for money and needed a new, more conventional job with regular hours. (Her previous fieldwork, studying the effects of wildfires on lizards in the chaparral, involved a lot of jumping around during controlled burns.) She was on the verge of moving back in with her mom in Austin when Mattoni took her on. She'd be breeding the butterfly in a fluorescent-lit double-wide trailer at the fuel depot.

The work wore Jana down. Every morning, she had to go through and individually inspect several thousand eggs or larvae with a dissecting microscope. It was horribly labor intensive—the larvae can be about the size of an eyelash—and, worse, extremely depressing: huge numbers of the young turned up dead each day; she was essentially a butterfly undertaker. ("I cried a lot," she told me.) Meanwhile, the captive rearing program was producing only enough butterflies to breed more butterflies in captivity, not enough to release any at the fuel depot, or to set up new populations elsewhere in the species' historic range. Jana found this discouraging. It didn't represent a failure, exactly, but an absence of hope. She talks about her work as "undoing an injustice that was done to nature by man." But here nothing was being done to correct the injustice done to the butterfly or to restore its former glory. The butterfly was still stifled—trapped on the military base. She felt stifled, too.

Mattoni, her boss, was an accomplished but, by all accounts, bullheaded man. Captive rearing of endangered butterflies has always been as much of an art as a science, with scientists developing their own idiosyncratic strategies and tricks. Mattoni was a pioneer of the field. In the late seventies, for example, he'd bred pink bollworm moths for the U.S. Department of Agriculture. (The moths Mattoni bred were sterilized, then released into the wild to mate with the pink bollworms that were plaguing California's cotton crop, eventually wearing that wild population down; it was a canny form of pest control.) Mattoni founded a company and worked up a large-scale, laboratory-like moth "factory" and, by 1982, was churning out two million pink bollworms a day. He became known in entomological circles for a lecture he gave called "How to Breed Two Million Moths for Fun and Profit." His approach with the Palos Verdes blues wasn't nearly as industrialized or clinical, but the general ethos was the same. He'd started on the project believing that there was no butterfly he couldn't breed by the millions, either, if he wanted to.

Jana, however, suspected that certain ways Mattoni insisted she handle the larvae were causing the high death rates. She wondered if she could do better by the butterfly. Eventually, she was entrusted with eighteen female Palos Verdes blues as guinea pigs for developing a new method. She gave these butterflies the same unbending attentiveness her toddler and newborn were demanding of her, throwing herself into re-engineering the entire breeding process. Within two years, the captive stock exploded. Soon she had enough butterflies to start a small new population on an oceanfront preserve, away from the fuel depot. It felt like a triumph. She got a tattoo of a Palos Verdes blue on her right ankle with the words ". . . And then she flew."

As her divorce plodded on and the legal documents piled up, Jana felt herself clinging to the Palos Verdes blue, identifying with it in a richly personal way that many scientists might not admit to—as two kindred underdogs, spurned but battling their way out of a corner. She wasn't just anthropomorphizing the butterfly; you could say she was Oprah-pomorphizing it. The butterfly was becoming her avatar, a gauge of her ability to reinvent and empower herself as a scientist and single mother. Resuscitating the Palos Verdes blue became both a literal test of her abilities and a metaphor for her own resilience. "That was me redefining myself," she told me. The symbolism was almost too easy. Butterflies have always been symbols of rebirth and renewal, and the closer Jana got, the more levels of metaphor she saw. A larva, for example, doesn't just develop into a butterfly inside the pupa; it first breaks down completely into an amorphous goop, then re-forms. Mattoni called it "the soup stage."

"You're not what you were before," Jana told me, "but neither are you what you're going to be. The soup stage really sucks, but you just have to embrace being soup for a while."

In 2007, Fish and Wildlife asked Jana to adapt her techniques to breed Lange's metalmarks, and she set up the Butterfly Project at Moorpark under the auspices of a nonprofit called the Urban Wildlands Group. By then, Jana had had a falling out with Mattoni, which she wasn't keen to discuss with me. All she would say is "Rudi knows more about butterflies than I ever will."

The ranks of California butterfly people are riddled with eccentrics. (I heard about a lepidopterist with a fantastically large collection of cocktail swizzle sticks, for example, and one who performs in a cowboy-costumed vaudevillian song-and-dance duo—come to think of it, I'm not sure they aren't actually the same guy.) But I was quickly getting the impression that Rudi Mattoni stands out. For decades, he'd been an idealistic and truculent crusader for endangered insects around Southern California, and seems to have left a trail of cheesed-off local governments, corporations, developers, and even other conservationists in his wake. A former colleague told me, "Rudi always had a bit of a persecution complex, like a prophet not recognized in his own time." Still, I gathered that his influence was enormous. His former partners and protégés seemed to have fanned out to work on butterfly conservation projects across the state. One senior Fish and Wildlife employee, who worked with Mattoni in the eighties, described him to me as "one of the last of the nineteenth-century naturalists."

Mattoni was born in 1927 and grew up in Beverly Hills, collecting butterflies and hunting for horned lizards. He'd been on the Lepidopterists' Society's trip to Antioch Dunes in 1954—the one that altered the course of Jerry Powell's career—but one of the seminal experiences of Mattoni's early life happened years earlier, around 1943, when Mattoni and a few butterfly-collecting friends ditched high school one February morning and drove out to a rocky riverbed east of Los Angeles known as the San Gabriel Wash. In an area no bigger than three square miles, they encountered thousands of small, powdery metallic blue butterflies called Sonoran blues, *Philotes sonorensis,* twinkling like tinsel in the air. The density of the population rivaled any other butterfly population in the world, and the sight of that many butterflies in one place, all streaming through their discordant orbits, was so breathtaking that Mattoni would still be able to close his eyes and see it as an old man. But in 1967, the Army Corps of Engineers bulldozed the habitat as part of a project to shore up the local water supply. The plants and butterflies were gone. Mattoni took it personally. "After that," he would later say, "I thought, 'Fuck all. I'm never going to see another butterfly.'"

He gave up butterflies for at least a decade. He worked in the munitions industry at North American Aviation. He led a NASA experiment to shoot *Salmonella* and *E. coli* into outer space. He coauthored a manual called "Sanitation and Personal Hygiene During Aerospace Missions." He made and lost a couple of million dollars trading bonds.

Eventually, Mattoni found his way back to butterflies. In the early eighties, his was one of the loudest voices in what one conservationist describes as L.A.'s "crazy butterfly fringe," mounting a battle to save the endangered El Segundo blue on the last remnant of its habitat: three hundred acres of dunes between the runways of Los Angeles International Airport and a Chevron refinery—yet another ragged refuge for insects. Mattoni worked at the dunes for ten years, studying and restoring the habitat and pulling weeds, while airliners throttled skyward just above his head. By the late nineties, he was shouting down the enemies of another rare Southern California bug: the Delhi Sands flower-loving fly, an inch-long fly with bulging Martian eyes.

Several cities and the National Association of Home Builders were suing the federal government to undo the fly's protections under the Endangered Species Act. Its presence on 365 acres of junked-up land had stalled development, and the insect was being held up as a symbol of how wrongheaded and ludicrous environmental protection was getting. (As late as 2011, politicians were still trying to get the fly removed from the endangered species list.) The fact that the Delhi Sands flower-loving fly had the misfortune of being named "fly" made it the perfect target. Outraged businessmen turned up at hearings with flyswatters. "It's hard to throw your support behind a maggot," one city attorney told the press. Mattoni responded by telling a reporter, "The stupidity of politicians is so mind-boggling."

By this time, one former colleague told me, Mattoni was becoming "apocalyptically pessimistic." He was squabbling more, and more bitterly, with his enemies and allies, including the nonprofit overseeing the Palos Verdes blue's recovery. When the group's director fired Mattoni, Mattoni simply packed up all the butterfly pupae from the fuel depot, put them in his car, and drove to his house. It was soon after that that Mattoni "imploded, exploded, decided to hit the reset button," as the colleague put it. In 2003, at the age of seventy-six, he abruptly stepped down from his job at UCLA. He went to Buenos Aires, where he still lives.

Among the many papers Mattoni published in his career, I found one about the butterflies he'd seen as a teenager at the San Gabriel Wash. In it, Mattoni used specimens he'd collected there as a young man to demonstrate that those blue butterflies were actually exceedingly different from other subspecies of *Philotes sonorensis*. They should be considered their own subspecies, he argued—if only now, after their extinction. In keeping with tradition, Mattoni got to name the new butterfly in his paper.

He called it *Philotes sonorensis extinctis*—the "Human Folly blue."

The name alone suggests that Mattoni was beginning to see humanity with the same skepticism and disappointment that had crept up on William Temple Hornaday, that the baselines were shifting beneath his feet. He ended the paper curtly, betraying how pointless he was coming to believe his work was. "Should curiosity of biological matters survive for future humans," Mattoni wrote, "this note may be useful."

THE FIRST TIME I took my family to Antioch Dunes was two days before Isla's second birthday. "Good God," my wife, Wandee, gasped when the gypsum plant came into view.

When we got out of the car, Wandee thought she saw something flutter past. "Is that a butterfly?" she said. Isla suddenly whipped around, grabbing at the air. After that, we decided to keep Isla strapped to one of our backs in a carrier, on the off chance that she managed to snatch a Lange's or accidentally stomp an endangered plant, both federal offenses under Section 9(a)(2)(B) of the Endangered Species Act. In that fragile environment, our little girl suddenly seemed to have the capacity of a Godzilla—a heedless destroyer, trundling between bushes as though they were Japanese office towers.

There was no shortage of butterflies in Isla's life. They spread their sequined wings on her favorite hoodie and flitted out of sticker books, winding up on the walls. By now, the wild animals were everywhere in our house—the geese on her quilt, the fawn on her wall. They seemed to be spontaneously generating, like a cuddly infestation, spreading through every storybook on her shelf. I read that one researcher, pulling a random sample of a hundred recent children's books, found only eleven that did not have animals in them. And what really struck me as strange was how often those critters have nothing to do with nature at all, but are only arbitrary stand-ins for people: the ungainly pig that yearns to be a figure skater; the squirrels that look disapprovingly at the bear who cannot stop biting her nails; a family of raccoons that bakes hamentashen for the family of beavers at Purim. It had all started to feel slightly insane, and I was hungry for an explanation. As Kierán Suckling, the executive director of the Center for Biological Diversity, had pointed out to me, "Right when someone is learning to be human, we surround them with animals."

Almost from birth, kids seem drawn to other creatures on their own. In psychology studies, children as young as six months try to get closer to, and provoke more physical contact with, actual dogs and cats than they do battery-operated imitations. Infants smile more at a living rabbit than at a toy. Even two-day-old babies have been shown to pay closer attention to "a dozen spotlights representing the joints and contours of a walking hen" than to a similar, randomly generated pattern of lights. It all provides evidence for what Harvard entomologist E. O. Wilson has dubbed "biophilia"—his theory that human beings are inherently attuned to other life forms. It's as though we have a deep well of attention set aside for animals, a powerful but uncategorized interest, waiting to be channeled into more cogent emotions, such as fascination or fear.* The attraction is so strong that a pair of psychologists felt compelled to assert in one scientific paper that a *lack* of interest in animals among children "can be normal," too.

Children fixate on animals in their imaginative lives also. They see animals in the inkblots of the Rorschach test twice as often as adults do. When, in 1955, a Tufts University psychologist went into a New Haven preschool and asked kids to tell her a story that they'd made up

---

* For example, children have been shown to acquire fear of spiders and snakes more quickly than fear of guns and other human-manufactured dangers. In this case, there's a logical, evolutionary basis for biophilia: if you are an immobile baby spending a lot of time on the ground, it pays to learn quickly to fear snakes, spiders, and rats. Fear of big predators doesn't kick in until after four years old, about when the first human kids would have begun roaming unaccompanied outside of their camps.

on the spot, between 65 and 80 percent of them told her a story about animals. Other research has found that 60 percent of the dreams that children have between the ages of three and five years old are about animals. But as kids grow up, the percentage of animal dreams goes down. By the time they're fourteen, it's only 20 percent. Similarly, fears of beasts like lions and sharks peak during preschool, then are gradually replaced by more human terrors, such as death, kidnapping, and not fitting in at school. I found a melancholic subtext to all this research—the way our world intrudes on, and then finally blots out, even the wildlife in children's heads.

Adults, meanwhile, have always tended to see kids and animals as vaguely equivalent, or at least more like each other than like us. "Children," Sigmund Freud wrote, "show no trace of the arrogance which urges adult civilized men to draw a hard-and-fast line between their own nature and that of all other animals. Children have no scruples over allowing animals to rank as their full equals." Kids begin life naked, unable to speak, and motivated by only instincts and urges. Like a pet dog, they need to be fed, housebroken, and taught to sleep through the night without howling. For Freud, this animalness was problematic: socializing children meant sculpting their wildness into humanity. But these days, it's easy to feel that society is the problematic force; we see it despoiling so much. And so, feeling that we are losing the wild everywhere, we're prone to romanticize our wild children the same way we romanticize wild animals. Maybe we keep giving animal stuff to kids because their imaginations innately brim with animals, but maybe it's the other way around. Maybe we long to see children and animals together—free creatures living in an innocence we've strayed from.

It's impossible to know. Most scientific research focused on kids and animals dissects children's relationships with their pets, not their abstract feelings about wildlife or the many secondhand images of it they encounter. In 1979, however, the Yale social ecologist Stephen Kellert and a Fish and Wildlife Service employee, Miriam O. Westervelt, interviewed kids at twenty-two schools in Connecticut, in grades two through eleven, to gauge their attitudes toward wild animals. As far as I can tell, it's the only study of its kind. What they discovered is an obvious but deflating truth: little kids *are* like animals, too consumed by their own interests to register much concern or compassion for other animals in the abstract.

Kids under the age of six especially "were found to be egocentric, domineering, and self-serving," Kellert later wrote. "Young children reveal little recognition or appreciation of the autonomous feelings and independence of animals" and "also express the greatest fear of the natural world." It was the younger kids, not the eighth- or eleventh-graders, who were more likely to believe that farmers should "kill all the foxes" if a particular fox eats their chickens; that it's okay to slaughter animals for fur coats; that most wild animals are "dangerous to people"; and that all poisonous animals, like rattlesnakes, "should be gotten rid of." It was the younger kids who were more likely to agree with the statement "It's silly when people love animals as much as they love people," whereas virtually none of the teenagers believed it was silly. Most second-graders agreed with the statement "If they found oil where wild animals lived, we would have to get the oil, even if it harmed the animals." Eleventh-graders overwhelmingly did not.

"Our society frequently romanticizes young children's attitudes toward animals," Kellert writes, "believing that they possess some special intuitive affinity for the natural world and that animals constitute for young people little friends or kindred spirits." But the data was clear: the younger the kids, the more "exploitative, harsh, and unfeeling" they were—the more their relationship to wildlife was based on the satisfaction of "short-term needs and anxiety toward the unknown." Older kids wanted to go camping in wildlife habitats; younger ones wanted "to stay where lots of other people were."

We like to imagine our children as miniature noble savages, moving peacefully and naked among the beasts—"the naturals," as the first colonists called the Indians. But they're more like the colonists: greedy, vindictive, wary, shortsighted, and firing panicky musket shots at any rustling in the woods.

It's not their fault. They are behaving like children.

WANDEE RIGGED Isla onto her back, and the three of us tramped into the dunes. We'd come to meet Jana Johnson and her students, who'd driven up from Moorpark to collect their Lange's metalmark foundresses for the year.

After weaving through some oak trees in the valley between the power line towers, we found Ken Osborne, a veteran lepidopterist who assists Jana in the field. Osborne, who has the twiggy gray beard and long hair of an Allman brother, was stalking a single Lange's on a buckwheat leaf. He motioned for us to stand still. "That's large enough to be a female," he said to himself quietly. "So, yeah, we're going to take that." He swooped at it with a deft flick of his wrist, and laid the net, with the butterfly inside, on the ground. Then he knelt, put on his reading glasses, and said, "Yeah, female." He transferred it into a vial and handed it to one of Jana's students.

Jana had set up an impromptu command center on a folding table at the end of the dirt road into the refuge. She seemed not exactly stressed but focused; she compares collecting foundresses to bringing patients into an ICU. They would be taking home four females again, and as each one was hustled in from the dunes in a vial, she and her students worked quickly to feed it honey-water with a Q-tip, transfer it onto a potted buckwheat plant, encase the top of the plant in a deli container and begin monitoring it for egg-laying. "This one's fat and really doesn't like being in the thing," one student observed.

Isla watched the whole collection procedure intently, from capture to coleslaw container. But she had no discernible reaction. (Frankly, she wouldn't seem to care much about the butterflies when I took her to the dunes the next summer, either, when she was three, though she did intently keep a lookout for Humphrey the Humpback that year, just in case the whale happened to pass down the river again—I'd been reading her a picture book about the rescue called *Humphrey the Lost Whale: A True Story*.) Clearly, Isla was interested in the Lange's metalmark only because the rest of us were so interested. She seemed to be puzzling out why such energy and attention were focused on these bugs, but was too shy to ask in front of the crowd of strangers. (Later, clicking through photos from that first trip, I would notice that, in every shot, Isla is looking not at the butterfly but at whoever was making the effort to point it out to her.) Ultimately, she spent most of the day down by the riverbank with Wandee, rambunctiously chucking sticks into the water, digging for snakes, and sticking her fists in the mud. Then, having worn herself out, she fell right asleep on the car ride home. I'd been flying around North America looking at wild animals. Suddenly, I realized that, all along, I had one of those at home.

Over the next year, though, I'd watch strains of her wildness begin to disappear. As two-year-old Isla became three-year-old Isla, they were replaced by irrepressible signs of personhood: a person who distinguished between Safeway and Trader Joe's and could swipe her way around an iPhone. One morning, when I took several minutes on the sidewalk to explain, rather elaborately, how a cement mixer worked, I could tell that most of what I'd just blabbered about was actually being absorbed and deciphered. She was becoming just a little bit reasonable, learning to scrunch her unruly urges into intelligible complaints and requests. She'd also stopped biting us—it had been a big problem—but if she was especially furious, or just frustrated, would still occasionally mash her lips against the leg of my jeans, straining hard to keep her teeth back.

It was astounding—but also slightly sad to watch. I felt the same pang that I imagined people felt in 1985, watching Humphrey come in from the wide-open ocean and bolt up the river, into the more limited, dirtier world the rest of us have to live in. Isla was becoming one of us, but she was losing something, too. Maybe it was only some false authenticity that I'd projected onto her. But every hint of its disappearance still bummed me out.

This is what was behind my urge to show Isla these animals in the first place, I realized. She'd come into the world as an animal. I watched it happen—it was magnificently, bloodily biological. I wanted to keep her feeling like a part of that expansive natural world, and not just of our self-contained human one. For me, wildlife has always been a reminder of all the mystery that exists outside my own experience— out there, beyond the suburban rec room I felt trapped in as a kid, watching *Wild America* on PBS. There's a special amazement that comes from watching a grizzly smack a salmon out of a river, or even from seeing just how hideous certain bottom-dwelling fish look. It enlarges your sense of the world, the way looking out from the top of a tall hill does. It's the perspective that William Temple Hornaday feared American kids would lose if they only stared into microscopes instead of strolling through the woods with a field notebook.

It was in Hornaday's time, after all, when this nostalgia for wilderness arguably first took hold in the United States. And even then, Americans were quick to cast nature as a refuge for our children, a way to escape the society we felt ambiguous about bringing them into. In 1909, the *New York Times* praised the shift in children's toys, from baby dolls to the new fad of stuffed animals, as working this same magic. The terrible thing about dolls was that they were little replicas of us, the *Times* wrote—they "concentrate a child's attention upon his own human qualities." Teddy bears did not. "And to get the child free from himself is the great modern problem."

Isla didn't need any reminders yet. She was still free from herself— still wild. But before long, she wouldn't be. Her toothbrush in the shape of an orca had first given way to one with Winnie-the-Pooh on the handle. By the time she turned four, she would have one emblazoned with Disney princesses. Soon, the research explained, the animals would be receding even from her dreams at night. Before that happened, I wanted to plant one of her actual feet in their world.

10.

THE SOUP STAGE

Satoshi Tajiri grew up in a town west of Tokyo called Machida. The Machida of his childhood, in the 1960s, was bucolic. There were ponds, gushing rivers, rice paddies, and quiet forests to roam. Tajiri spent his youth prowling that countryside, observing and collecting insects. He dreamed of being an entomologist. "Every new insect was a wonderful mystery. And as I searched for more, I would find more," he later remembered. His friends called him "Dr. Bug."

But by the late seventies, with metropolitan Tokyo bulging outward, the town of Machida had changed. Railway lines and roads slashed through the landscape. Apartment buildings and shopping malls appeared. Insects were harder to find. "The change was so dramatic. A fishing pond would become an arcade center," Tajiri said.

He adapted. As insect habitats dwindled, he funneled that attention into the video game arcades that replaced them. He became mesmerized by Space Invaders. He dismantled a Nintendo to see how it worked and learned to program his own games. By the time he was an adult, Tajiri said, "Places to catch insects [were] rare because of urbanization. Kids play in their homes now, and a lot had forgotten about catching insects. So had I." So, in the nineties, he designed a Nintendo game that tapped into his childhood impulse for bug hunting—a virtual world, bursting with fictional biodiversity. It now contains more than 640 precisely named "species" of critters, all of them waiting to be collected and traded with friends. Tajiri's game is Pokémon.

The Pokémon franchise has evolved into films, television shows, and comic books. Most conspicuously, it's evolved into Pokémon trading cards, letting kids physically sort and classify the creatures they collect, just as kids once organized and reorganized species in their bug collections.

It's often said that kids are born taxonomists. There's a certain satisfaction to making sense of the world by chopping it up into increasingly specific categories. It was a love of collecting and identifying insects at an early age that had sent nearly all the lepidopterists I was encountering into conservation—it *was* the experience that the lepidopterist Robert Michael Pyle was talking about when he talked about the extinction of experience. Now, however, one study found that a typical eight-year-old in Britain can identify upward of 120 different Pokémon species, but only fifty different real plant and animal species native to his or her area, like oak trees or badgers. It's the same great modern problem the *New York Times* had identified a hundred years ago: kids are being pulled away from nature and inward—which is to say, in the exact opposite direction I wanted to nudge Isla. Slowly, children's innate flair for taxonomy—the science of knowing, naming, and ordering the many forms of life on Earth—is being transferred away from actual wilderness and applied to a world that Tajiri invented as a consolation prize for its loss.

I spent much of the winter between my family's two trips to Antioch Dunes learning about taxonomy. The subject had kept coming up, but, frankly, all those Latin names of butterfly species and subspecies had seemed too tedious to matter. I was wrong. It turns out that taxonomy complicates the story of the Lange's metalmark, and maybe of any conservation, as powerfully as the problem of shifting baselines does. It would lead me into my own soup stage that winter. What I knew broke down into a writhing mush.

ALL TAXONOMY STARTS IN IGNORANCE. To infants, every four-legged animal is a "doggie." But kids gradually cue into subtler variations in size and shape, and features like stripes, manes, and snouts. All those doggies come into sharper focus as zebras, polar bears, and goats.

In science, the process is the same. Organisms are named and grouped based on a small number of obvious differences. But the more closely you look at those animals, the more possibilities for difference you pick up on. And, as a result, the more differences you'll see. You refine your categorization of animals just by noticing that new categories exist: the shape of a tail, the number of teeth, all the way down to minuscule variations in the splotches and colors on butterflies' wings. Taxonomy is the science—or maybe just the art—of deciding which of those distinctions are ultimately worth making: where to split the spectrum of nature into freestanding groups.

The theory of evolution complicated things for taxonomists. It revealed that a species isn't an unchanging type but a snapshot in a long and unpredictable process. Suddenly, it became less obvious where taxonomists were supposed to draw their lines. In 1942, one evolutionary biologist proposed the "biological species concept," defining a species as a group of organisms that breed with each other and not with outsiders. This had a simplifying effect, sweeping creatures that had been separate species back into larger groups. But the human instinct to make distinctions still needed an outlet. Taxonomists simply took all those different-seeming things that no longer qualified as species and called them subspecies instead. They started picking apart other species, too, establishing new subspecies. It got out of hand. The reckless proliferation of subspecies reflected, one critic wrote, "an undesirable trend toward taxonomic chaos." Eventually, most fields reined in the problem. But lepidopterists tended to keep on naming and dividing at will. "Butterfly people have nothing else to do," one veteran lepidopterist told me. "There are far fewer butterfly species than there are butterfly enthusiasts. So they love to come up with new names. They say, 'Oh, they look smaller and darker here. So let's give it a name.'"

For butterfly people, the key to identifying a new subspecies has always been "good eyes"—picking out visual differences in size and in the patterns and colors on wings. There have always been rules floating around to standardize the process, but not especially functional rules. As early as 1953, scientists recognized that subspecies were "inherently subjective and even arbitrary." They are human projections—matters of opinions with no agreed-upon empirical measures to check those opinions against. "A subspecies is anything that anybody who can wield a pen and get a paper published says it is," another veteran of the field says.

As a result, fights about the uniqueness of certain species and subspecies are common. As in many fields, lepidopterists divide into two camps: "lumpers," who are comfortable gathering up large groups of different-looking butterflies under the same species or subspecies, and "splitters," who prefer more painstaking divisions. (I found the antipathy between lumpers and splitters to be pretty shocking. Sometimes, profanity gets thrown around.) Quarrels over taxonomy sound geeky, like tiny feuds between the persnickety and slightly less persnickety. But they occasionally have broad consequences—namely when, via the Endangered Species Act, the federal government is deciding whether to throw its weight into saving one of the creatures being quarreled about.

In an exhaustive essay on this issue, the environmental legal scholar Holly Doremus describes the Endangered Species Act as imposing a "static vision" on a "dynamic world." The law is forced to fix the ambiguities of nature, like the blurry borders between species, into the binding certainty of legalese. The act does define "species" flexibly. (Subspecies can be listed as an endangered "species," for example.) But even so, there's often disagreement about how unique a critter actually is—and, consequently, whether saving it is worth redirecting public money and rewriting land use policies. There's often even disagreement about how that uniqueness should be measured and judged. In 2005, for example, the Fish and Wildlife Service was asked to decide whether a particular mouse in Colorado and Wyoming, the Preble's meadow jumping mouse, was a discrete subspecies—and therefore rare and in need of protection—or merely part of another, widespread subspecies of mouse. The government got conflicting opinions from fourteen different scientists. Two scientists had done DNA analysis, showing the level of genetic difference between the two mice, but still reached the exact opposite conclusion. This was because there was no consensus about how different the two genomes needed to be for the mice to qualify as two different subspecies.

To make things even messier, we are also quietly accelerating evolutionary change. In Europe, for example, certain songbirds have forked into different rural and urban species, each uniquely adapting to the habitat we've built around it. Around the world, all kinds of species are now shrinking—their average body size is getting smaller—because generations of human hunters have removed the biggest, fittest animals from their gene pools. And climate change, by warming up the Arctic, is allowing grizzlies to range farther north, where they and polar bears have started interbreeding. Eventually, the polar bear may go extinct only after being absorbed into a new and unrecognizable hybrid species, which scientists, for the time being, can't decide whether to call pizzlies or grolars.

That is, we are not just identifying uniqueness but *creating* uniqueness—uniqueness that, one day, we might be similarly impressed with and feel obligated to protect. We imagine conservation as keeping essential shapes of nature locked in place. But those shapes are sometimes just projections, a zone of ever-shifting fuzziness that we've chosen to draw a solid line around.

WHICH BRINGS me back to the Lange's metalmark butterfly.

In fact, it was while talking about the butterfly to Jerry Powell in his office at Berkeley one afternoon that I was first tugged into this taxonomic quicksand. Powell was grousing about the Fish and Wildlife Service's effort to save the Lange's—all the weed pulling and captive breeding, which, after all his years of studying the Antioch Dunes, seemed to him both futile and meaningless. Finally, I asked if there was a chance that another population of Lange's metalmarks might be discovered outside the dunes one day, just as Rudi Mattoni had rediscovered the Palos Verdes blue at the fuel depot. Powell considered my question. "Not strictly speaking," he said. But, he added, it depends on what you mean by Lange's metalmarks.

The Lange's is a subspecies of the Mormon metalmark species. Like all butterfly species and subspecies, it is defined in the scientific literature as a butterfly that fits a particular description—in this case, a physical description meticulously detailed from the fifty specimens that Harry Lange first collected at Antioch Dunes in the 1930s. ("Fringes" of the wings are "checkered black and white, the black disposed at the ends of the nervules and veins. Outer third of wing [upper surface of primaries], black, with a submarginal row of seven small white spots, all of about equal size, and a second row of larger subtriangulate white spots internal to the first, which are of unequal size, the third and sixth being the largest." And so on—the written description is two pages long.) Other kinds of Mormon metalmarks, with different wings, fly throughout the western United States.

Powell explained that, years ago, he'd found a large population of Mormon metalmarks outside the town of Mendota, about 120 miles southeast of Antioch Dunes. Astoundingly, one-third of these butterflies looked exactly like Lange's metalmarks. They were indistinguishable from the butterfly described in the literature.

The existence of these butterflies raises some awkward philosophical questions. You could argue, for example, that these particular butterflies *are* Lange's metalmarks—because they fit the description. But this logic leads to some very illogical places. It would mean that certain butterflies in Mendota, with certain wing patterns, would be federally protected as endangered species, while the other two-thirds of the butterflies they fly with—including, sometimes, their own brothers and sisters—wouldn't qualify for protection. On the other hand, you could also argue that Powell's discovery meant that the Lange's metalmark isn't actually endangered anymore, or even rare. We could stop worrying about the butterflies at Antioch Dunes and let them die, knowing there was a healthy sanctuary of "Lange's metalmarks" in Mendota. This would be phenomenally good news—a relief. But it doesn't feel that way. It feels like an accounting trick, the shifting of a few insects from one column of a ledger to another to mask a bad investment.

Powell told me he was now sending various Mormon metalmark specimens to colleagues at the University of Alberta for DNA analysis, and so, later that winter, I called them up. A graduate student, Benjamin Proshek, had compared the mitochondrial DNA of the Lange's and the Mendota butterflies and found that, ultimately, they are not very genetically similar after all. The identical markings on the two butterflies' wings was likely just a coincidence, Proshek told me—two packets of different genes happened to produce the same superficial result. I took this to mean that the Lange's, in some deeply scientific way, was still unique. I felt the claustrophobic and contentious world of butterfly politics settle back on its axis.

But Proshek wasn't done. He also told me that he compared the Lange's with its closest neighboring metalmark, a population of butterflies that lives on Mount Diablo, ten miles away from Antioch Dunes. To a trained eye, the wings of this subspecies look incredibly different from the Lange's. But genetically, they were actually much more closely related to Lange's than the Mendota metalmarks; the difference in their mitochondrial DNA was less than half a percent, Proshek said—extremely slim. He suspected that pioneers from Mount Diablo must have colonized the Antioch Dunes six to ten thousand years ago. These particular butterflies were probably oddballs: by chance, they had a slightly higher proportion of certain gene variants than others, compared to the rest of the population. Once isolated in Antioch, they began inbreeding, exaggerating those differences, and their offsprings' appearance morphed into that of the modern Lange's. Slowly, traits were expressed that had been hiding in the genome of the Mount Diablo butterflies all along—and are still hiding there.

This genetic variability hidden within individual butterflies happened to be one of Rudi Mattoni's many fixations. In the early eighties, Mattoni captured a single Gulf fritillary butterfly in Southern California, a subspecies whose wings are reddish orange with sparse black embellishments. With a colleague, he reared its eggs and began selectively inbreeding the offspring. They paired together the butterflies with the most pronounced black markings on their wings, and also paired together the ones with the least black on their wings. After seven generations, Mattoni and his colleague had produced a profusion of astoundingly different-looking butterflies and seemed to be on their way to creating two distinct races: one with totally black wings and one with totally orange wings. The array of oddities and rogues they'd wrung from a single female would have been easily mistaken for "several different subspecies, if not species" of butterfly if discovered in the wild, they wrote. The experiment challenged the meaningfulness of defining subspecies by their appearance. It also challenged the meaningfulness of conservation based around those definitions. After all, it might be possible to capture Mormon metalmarks from Mount Diablo and, through this same kind of selective inbreeding, conjure butterflies that look exactly like Lange's metalmarks all over again. This happened naturally six or ten thousand years ago, but it could, hypothetically, be done artificially, too—tweaking the butterflies' appearance over several generations to meet the description of Lange's in the literature, the same way American Kennel Club breeders refine the noses and coats of purebred dogs to meet the written descriptions of each breed.

It begged the question: what, by saving the Lange's metalmark butterfly, would you actually save? Maybe it's important to preserve a butterfly that looks a certain way. But there are identical-looking butterflies in Mendota. Maybe it's about saving particular genes. But those genes are likely still up on Mount Diablo. Maybe it's simply about staving off the extinction of experience, of keeping a creature and its context intact. And yet the experience of the Antioch Dunes—Jerry Powell's experience of them, anyway—had gone extinct many decades ago.

I've sat and thought about all of this for a long, long time. Frankly, I don't see any upshot—except simply to acknowledge how very disturbing it feels. Whatever principle was driving everyone to save the Lange's metalmark felt so fragile, so washed in emotion. My instinct had been to try to lash it back to some immovable, empirical anchor. But now it wasn't clear what all these scientific answers I'd collected added up to. Maybe, in the end, those emotions are all we have to go on—because when you looked at the butterfly rationally, nothing made intuitive sense.

I DID SOMETHING else unsettling that winter. I talked to Rudi Mattoni.

Mattoni still lives in Buenos Aires. He responded to my initial e-mail, explaining that he'd gotten involved with an artists' colony down there and sending me a video he'd made so that I would understand where he was coming from.

The video began with a tutorial about biodiversity and slides detailing the extinction of various species. "Why should I give a damn?" another slide asked, and then the presentation scattered into a succession of stock images: a clear-cut forest, smokestacks, a human hand holding a globe on fire, a pile of rats, starving African children, a mushroom cloud. Soon there were photos of an art opening Mattoni had organized at a gallery in Buenos Aires. There were watercolors; a sculpture of butterflies and warplanes; photos of dead butterflies with their wings battened open with screws and nuts. The video might have felt chaotic, except that the entire time a lilting big band tune played in the background: a crackling 1944 recording of Benny Goodman and his orchestra doing "Poor Butterfly," on repeat, with polite applause sounding at the end of each loop. "The sixth extinction and the end of Nature," one slide read near the end. "We now live in a manmade world."

Mattoni and I made plans to talk on Skype, and a few days later, he appeared as a grainy image on my screen. His black hair was streaked with white, and he hunched out of a dark room toward his computer, peering over his glasses. He turned out of frame to tell a housecat named Trotsky to scram off the top of his printer, and settled in.

Mattoni first wanted to make sure that I'd read the text of the Endangered Species Act. There's a clause right near the top that nobody remembers, he said. "And it's the whole soul of the Endangered Species Act." It begins: "The purposes of this Act are to provide a means whereby the ecosystems upon which endangered and threatened species depend may be conserved."

It's the *ecosystems,* Mattoni said—the ecosystems are the point. It's an article of faith among conservationists that ordinary people's affection for particular charismatic animals can be widened into concern about the deteriorating natural systems they rely on, as with the polar bear and climate change. But Mattoni's experience had convinced him this was impossible. After hammering away with that strategy for decades in California, "I never got it through anyone's head." The butterflies were just a means to preserve wild places, but all the attention got lavished on the butterflies themselves. "People would say, 'How do we save the blue butterflies?'" Mattoni told me. "And I'd say, 'I don't give a shit about the blue butterflies.'"

Mattoni hadn't heard that the Lange's metalmark was being bred in captivity now, and when I told him why—that peak counts had dropped into the double digits—he was thrown. "Jesus! What happened?" he said. The news only confirmed his thesis. In Antioch, a gorgeous butterfly had been singled out and perpetuated, just barely, decades after the entire ecological context around it had come undone. "Once the habitat is gone, it's gone," Mattoni said. "It's too complex— you can't put these things back." He saw captive rearing as "basically trivial," which is how he saw a lot of conservation. "It's all theater," he said repeatedly. "I've come to the conclusion that conservation is really kind of a dead cause. I think virtually all of these efforts are worthless.

With climate change, Christ knows what's going to happen. You can't even predict it. There's nothing you can do—nothing will stop this. Nothing will stop this until it all comes crashing down."

Scientists have failed, Mattoni told me. None of the congressional testimonies and interviews with the press, including his own, convinced anyone. Curating the art exhibit was his way of trying something different. But the exhibit wasn't meant to inspire people to preserve biodiversity, as I took it to be. The exhibit was meant to communicate the need to *catalog* biodiversity, he said. Mattoni wanted to see a full-scale effort to collect and describe all the earth's plants and animals so that we'd at least have a record of what we destroyed. He was about to dedicate his life to this work. He'd just bought thirty acres along the Rio de la Plata in Uruguay—one of the few fairly natural remnants of a forest cut down a long time ago. There were big fish in his stream, he said, and wild boar running around; he could live on that land happily for the rest of his life. His plan was to do a scrupulous survey of every insect living there—for as long as he could still work, even if it meant netting bugs from a wheelchair.

Mattoni had traveled on exactly the opposite path as William Temple Hornaday. His vision had degenerated from trying to save species in the wild to giving them a fitting memorial. Hornaday had started by stuffing buffalo; Mattoni would end by mounting dead bugs. "We're going to lose a lot, and nothing's going to stop it," he told me. "But the unforgivable sin is, we don't even know what's here. I don't know what else to do, Jon. I think you can see it: I've given up."

I WENT to one last butterfly count at Antioch Dunes. It was a brilliant, gusty morning at the end of August, a year after my first butterfly count, and by that time I'd gotten fairly good at appreciating the refuge for the eclectically weedy and artificial place that it was.

Across the river, the white wind turbines tumbled like modern dancers, and the gypsum plant's sky-blue water tower stood in faint relief
to the actual sky. Way up at the foot of the western utility tower, one
side of a long, flat field had been transformed into an utter jungle by
a lurid green, leafy invasive plant from Asia called tree of heaven—
largely chopped down to its stumps, but still stubbornly booming in
places. Though it was a war zone in ecological terms, it felt hauntingly
gorgeous as we formed a long line and strolled through.

I came to count butterflies with Liam O'Brien again. And not
long after all the volunteers had familiarized themselves with the laminated butterfly mug shots, zeroed out their clicker counters, and
started prowling through the underbrush, Liam was doing his thing,
shouting, "Did everybody see the pygmy blue?" and kneeling by a
plume of Russian thistle to make sure that no one missed out on seeing "the smallest butterfly in the U.S.!" Russian thistle, which the
pygmy blue eats, is an invasive plant—it didn't belong at the dunes.
But you could argue that it brought the butterfly with it, contributing
to the experience of the property in its own way. "This one's a female,"
Liam said, treating us to one of his theatrical little lectures. "The
male's *half* that size!"

No Lange's metalmarks had been seen yet that summer, even after
weeks of counts. It was worrisome. Liam later told me that, walking
around, he kept thinking, even if they still were here, how difficult it
was going to be for the males and females to find each other and mate,
with so few butterflies left in such a relatively big space. It seemed entirely possible, too, that we might believe the species went extinct
when it had not—that whole summers could go by in which the small
cluster of inexperienced butterfly counters walking the dunes twice a
week never stumbled into any of the exceedingly small number of butterflies flying there; that not a single Lange's would even register as a
disruption in anyone's peripheral vision; that the humans and the but-

terflies could just keep orbiting one another obliviously, like separate worlds.

Ultimately, it was Liam who spotted the first Lange's metalmark of the season that day—and as soon as he did, Louis Terrazas and the other Fish and Wildlife staff on-site began texting and phoning their superiors to report the good news: the species had survived the winter. Then, as our group started climbing one of the utility tower hills, a man named Brent Plater thought he saw another and shouted, "I got one with its wings open here!" But then Plater's voice crumpled a little and he said, "Nope, it doesn't look right," so we moved on.

Plater is a young attorney who radiates a low-key sincerity and narrows his eyes when he thinks. He directs a small nonprofit in San Francisco called the Wild Equity Institute, and that year was preparing a federal lawsuit to help the Lange's. The suit sought to block three power generating stations just down the road: two new natural gas plants that were in construction, and an older plant that was operating with a lapsed permit. Power plants emit nitrogen—the main chemical in fertilizer—which settles in the surrounding soil, altering its chemistry, and boosts the growth of certain vegetation. The level of nitrogen around Antioch Dunes was already abnormally high, and new research was identifying high nitrogen concentrations as a severe threat to imperiled butterflies, since nitrogen often gives an even bigger boost to the invasive plants that their host plants are losing ground to. The U.S. Fish and Wildlife Service had recently written to the California Energy Commission, warning that the new power plants were "virtually certain" to harm the Lange's, and likely to kill it off completely.

Plater's lawsuit had the outlines of a stereotypical endangered species fight—a butterfly versus energy and jobs. A couple of local right-wing blogs were already belittling the suit and the bug, and many residents of Antioch and neighboring towns supported the power plants—they would be huge moneymakers for the community. But a lot was getting lost in the coverage, including why the power plants were clustered around Antioch to begin with. In part, it was because nearby San Francisco, which was demanding more energy, was also trying to go green and had made it all but impossible for new power plants to be built within its own city limits.

The problem felt intractable; everyone seemed to have a legitimate complaint. I wondered aloud to Brent Plater if the Lange's wasn't, in some more meaningful way, already lost. Its entire context was gone. Why not just let civilization have its power plants this time?

Plater told me that if humanity agreed to set aside some significant percentage of the earth for other species, then he'd be willing to make those kinds of concessions—to lose the Lange's. But the balance is so out of whack that every battle is now a battle of principle that can't be forfeited. Realistically, Plater conceded, he didn't know if he could stop the power plants. But he hoped to compel a settlement. Potentially, some serious money could be squeezed out of the companies that owned the plants—enough to finance a much more comprehensive restoration of the Antioch Dunes than anything the government had been capable of so far. Maybe they could even truck in some phenomenal amount of sand, dump it everywhere, and try to start over. In short, the lawsuit could be yet another freak turning point in the story of the Lange's metalmark: another forty-ton whale surfacing out of nowhere, but this time to restore order rather than exacerbate the chaos.

It was something to hope for. Frankly, it was all starting to feel a little pathological otherwise—this grasping after a butterfly that only got harder to pin down the closer you got. When we look at nature, maybe most of what we see is lines that we've superimposed there ourselves: taxonomic lines, legal lines, baselines of how we believe the world is supposed to look. These lines have only as much authority as we give them. It's our emotions that fasten them in place: how deeply we *believe* in their truths, and how guilty or queasy we feel when we come close to crossing them, even when crossing them may be the rational thing to do.

Maybe the most imposing line is the one we imagine between ourselves and nature—the belief that there is such a thing as pristine nature, and that it is sacrosanct, and that any changes we trigger in it can only be disfiguring. Recently, small pockets of conservationists have challenged that idea, arguing instead, just as the proposal for Pleistocene Rewilding did, that we should be actively seeking to cultivate a *new* nature, instead of struggling to forestall the disintegration of the one we happened to inherit. There's talk about the "managed relocation" of species—picking up animal populations and airlifting them into new habitats where they're more likely to survive climate change. There are calls to stop the blanket vilification of invasive plants, to accept that weeds are not going away, and to realize that they can be parts of equally biodiverse "blended" ecosystems rather than only blights on the "native" ecosystems we feel such nostalgia for. Seizing that kind of creative freedom—owning up to our power on Earth and exerting it—is either inspiring or existentially terrifying, depending on whom you ask.

In Antioch, meanwhile, people were clinging to the last Lange's metalmarks—believing in the butterfly and clapping as hard as they could, so that, like Tinker Bell, the species wouldn't disappear from the stage. But what if the greater, more progressive challenge was to work through the guilt and knowingly let the butterfly go?

In the end, part of me wants to argue for that. But, then again, maybe letting go once only leads to more letting go. Maybe you have to believe in the value of everything to believe in the value of anything. Maybe giving in a little only hastens the terminal disenchantment I'd seen afflicting all those old men—the many Rudi Mattonis who tried to hold on to so many things they believed in and, looking back on their lives, believed they'd lost.

I felt stuck, in other words. And I wondered if I wasn't the only one. The week after the butterfly count, I heard that peak count reached twenty-four and still seemed to be rising. The previous year's peak was twenty-eight, and Louis Terrazas told me, "If we get more than last year, I'm definitely going to throw a party or something." In the end, the count topped out at thirty-two that summer. It was an improvement—technically. But instead of a party, there was only another emergency meeting.

PART THREE

BIRDS

11.

CONSTRUCTION WORKERS

By the late 1940s, there were only about thirty whooping cranes left on Earth, and, for better or worse, America was about to level the clobbering weight of its attention on two of them.

Their names were Pete and Josephine, and they had been paired up at the Audubon Zoo in New Orleans so that they might breed. Pete and Josephine were not breeding, though, probably because the zoo's spectacularly ignorant superintendent, George Douglass, had confined the towering, long-legged birds to a pen hardly big enough for ducks, then stashed them in a noisy backyard, near the garbage, where they had rats to contend with, little shade to escape the heat, and only washtubs of dirty water to drink from—all of it hewing away at the whooping cranes' physical and mental health.

Pete, the male bird, seemed especially anxious—"hysterical," one visitor said. The crane had been a member of what was, by then, the only remaining wild flock of whooping cranes in the world. But in 1936, it was shot in the wing and left for dead while making one of its cohort's annual migrations between the Gulf Coast of Texas and a remote region of Alberta, Canada. Young sisters in Nebraska, out riding their bicycles, spotted Pete downed in a field and fetched their father, who bagged him. For the next twelve years, Pete would serve as a mascot for a local gun club. He could not fly and was missing one eye.

The crane that Pete had been shipped to New Orleans to mate with, Josephine, was the sole survivor of an encampment of whooping cranes that had once lived in southwestern Louisiana. One old-timer would remember, "It was beautiful to see them up there in the sky, always seven or eight in a bunch, circling and crossing each other like people square dancing." But by 1941, the Louisiana flock had almost entirely petered out. That fall, Josephine was discovered injured in a rice field and delivered to the zoo in New Orleans. That left only twenty-two whooping cranes in the wild. It would prove to be the all-time low for the species. Some people claim it was worse, that there were actually only fifteen whooping cranes left. Or fourteen. Others say twenty-one. It's one of many details, I would discover, that people who love the whooping crane can't agree on.

In fact, one writer notes, it wasn't long before Pete and Josephine's predicament in New Orleans incited a "head-on collision" of "two groups of people, both passionate about whooping cranes and passionately opposed to each other on the question of how to preserve them." The fury kicked off in 1948, when Robert Porter Allen visited the two cranes at the zoo. Allen was an ornithologist who was going to tremendous lengths to revive the species in the wild. Seeing the conditions there, Allen pleaded with George Douglass at least to feed the whooping cranes live blue crabs, which is what whooping cranes like to eat in the wild. Douglass had no background in animals whatsoever, and had been put in charge of the zoo through cronyism, it seems, after dropping out of college and working for his father at the city's waterworks. He asked Allen if blue crabs were "anything like" soft-shell crabs. Then, while Allen stood there horrified, Douglass phoned a local restaurant and ordered some for the birds. In an outraged letter to officials in Washington, Allen described the two whooping cranes as not living, but "merely existing." "The species has suffered much," he wrote, "but those two birds represent the lowest rung. I could have wept when I saw them, if I hadn't been so numb with anger and shame."

The whooping crane stands five feet tall, with a wingspan as long as eight feet. It is the tallest North American bird—a sixty-million-year-old species and one of the few living relics of all that vanished, Pleistocene-era megafauna. As salivating dire wolves and two-hundred-pound beavers plodded around America, the whooping crane stood poised in shallow wetlands, playing the long game. The bird is reclusive and headstrong, and believed to require a full square mile around its nest all to itself. So, as humans crowded into North America, then started draining the swamps where it lived for agriculture, the whooping crane couldn't easily adapt. The bird's size, elusiveness, and pure white feathers also made it a satisfying target for hunters—as did its abrasive bugling, which can carry for several miles. The species was rumored to be extinct many times. A 1904 *Washington Post* headline read: "Two Nebraska Hunters Kill the Last of the Pompous Bird."

Before European settlement, more than ten thousand whooping cranes lived across North America, from Utah to New Jersey. But the die-off of the Louisiana population in the 1940s left only one wild flock—the twenty-two birds, or however many, that wintered on the marshes in Texas and flew to Canada every spring to breed. As the number of cranes returning to Texas every fall dipped and rose slightly, the tally was reported in major newspapers around the country. Like William Temple Hornaday's buffalo a few generations earlier, the whooping crane gained an anxious mystique. It was a wonderful animal that would soon be gone and, because it would soon be gone, it seemed more wonderful. The bird's wildness—its majestic ambivalence; the insolence, even, which it showed the human race—became its signature virtue, the thing that people who celebrate whooping cranes still celebrate the most. "This is a bird," Robert Porter Allen wrote, "that cannot compromise or adjust its way of life to ours. Could not by its very nature; could not even if we had allowed it the opportunity, which we did not. . . . Without meekness, without a sign of humility, it has refused to accept our idea of what the World should be like."

It wasn't easy, but Allen eventually wrestled Pete and Josephine away from George Douglass. The cranes were brought to a large outdoor enclosure at the Aransas National Wildlife Refuge in Texas, the marshes where the last wild flock spent the winter. There the situation improved—sort of. In the summer of 1949, Josephine laid two eggs. They were the first whooping crane eggs ever laid in captivity. But a few days later, Josephine and Pete inexplicably smashed both eggs to pieces. Then Pete died. Before long, George Douglass got in a truck, drove to Aransas, walked into the refuge manager's office, and said, "I've come to get my bird." Then he stuffed Josephine in the truck, along with her new mate, a male crane named Crip, and drove them back to New Orleans.

In New Orleans, Josephine eventually laid another egg. But Douglass, no doubt feeling that he'd performed a miracle and wanting credit for it, brought in a rabble of reporters and photographers. One of the men jabbed a stick into the cage—he thought it would be neat to taunt the cranes, apparently—and Crip, defending his nest, accidentally trampled the egg and destroyed it.

At that point, there were twenty-one whooping cranes on Earth. The situation looked hopeless. The *New York Times* blamed the species' looming extinction on its own "lack of cooperation."

SINCE THEN, generations of conservationists and government employees have collaborated to protect that last remaining wild flock of whooping cranes. The handful of birds that were left in the 1940s have rebounded into a population of about 275 today. This stands as one of the most extraordinary turnarounds in conservation history, and one of the most backbreaking. The cranes' winter and summer habitats in Texas and Alberta have been shielded from all kinds of intrusions and threats. Disputes have metastasized into lawsuits. Individual cranes, found injured along the migratory route, have been airlifted to veterinarians, and during droughts in Texas, the government has fed the cranes corn, apparently settling for itself all the same philosophical questions that I'd watched calls to feed polar bears raise in Churchill, after the video of those starving cubs got out.

Watching the effort to keep the Lange's metalmark flying at Antioch Dunes, I sometimes wondered what that management would look like in thirty or fifty years. The ecology around the butterfly—all the natural machinery supporting it—was basically gone. Still, it seemed likely that the Lange's could keep coasting along at its current listless clip indefinitely, as long as we stood behind it, pushing.

Turning to the whooping crane is like time-traveling nearly a century into that future. Once, in the days of Pete and Josephine, we'd backed the ancient bird into the same sort of squalid and discouraging corner that the butterfly is in now. But then we kept it from dying out. And we *kept* keeping it from dying out. And as each generation lessened, but never solved, the crane's many problems, then entrusted those problems to the next generation, we avoided the permanent dead end of its extinction—and as long as we avoided that, we kept alive all other possibilities for the bird.

The whooping crane has been the beneficiary of such intensive, progressively more bizarre and intimate human intervention that, even by the time it was listed as one of the first protected species under a precursor to the modern Endangered Species Act, in 1967, it was already a recognizable symbol of the greater cause of conservation. Now those working on the latest phase of the recovery are building creatively on that long legacy of hard work. And the crane is beginning to symbolize something else: the reaches of what's possible—how, even when such effort looks foolish or futile through the pinhole of the present, its value in the long term is unknowable.

As A KID, Joe Duff couldn't have guessed that the whooping crane would make it this far—and he never imagined that he'd be working to help the bird, either. Joe was born in 1949, the year when Pete and Josephine smashed their eggs at Aransas. He grew up in Ontario, Canada, and remembers hearing about the whooping crane his entire childhood. One day, his mother insisted there was a whooping crane by the creek behind their house. But Joe had just read an article about whooping cranes in *National Geographic* and remembers thinking that, with only twenty of these things in the whole world, one couldn't possibly be behind his house in Ontario. (In fact, one wasn't behind his house: the bird turned out to be a heron—it wasn't even white— but Joe wasn't in the habit of telling his mother she was wrong.) Joe liked birds and nature as a kid. But what he really loved was cars. And also art. By the time he was thirty-three, he owned his own commercial photography studio in downtown Toronto and specialized in photographing automobiles for ads and dealership brochures. He shot foreign luxury cars on the winding roads of Big Sur and commanded a rate of $2,000 or $3,000 a day. He went to a lot of parties. When he needed a root canal on one of his front teeth, he convinced a dentist to give him a cap with a small copper star inlaid on the front—still visible when he smiles. The star seemed flashy and hip at the time, but now Joe, at age sixty-one, brushes off the inevitable questions about it by saying, a little embarrassedly, "It's from a former life."

The other thing that Joe did in those days was fall in love with ultralights—a kind of very tiny airplane. An ultralight, sometimes called a trike, has an open cockpit and looks like a little amusement-park race car dangling from the wing of a hang glider. It's powered by a massive propeller in the back. The pilot steers by shifting a bar under the wing with his arms. Ultralights were not especially safe back then, and still aren't, to some extent. But flying them was exhilarating and attracted a subculture of sometimes reckless or mildly self-destructive pilots. It was frivolous, maybe. But it was under the wing of an ultralight that Joe's life sailed into its improbably altruistic second act, and into the realm of the whooping crane. "Now," he told me, "I can't think of anything more useless than the car catalogs I used to produce."

Joe now heads a nonprofit called Operation Migration, part of a coalition of amateurs, biologists, and government agencies that has been trying, for more than a decade, to create a second population of wild whooping cranes in America, basically from scratch. Maybe it's a sign of how low the bar has sunk in conservation, but even one of the field's most celebrated successes—the resurgence of that western flock of whooping cranes, from fifteen to almost three hundred birds—still doesn't give solid odds for the species' long-term survival. And so conservationists are now breeding birds in captivity, releasing them in Wisconsin, and laboring to get them migrating to Florida along a flyway that whooping cranes haven't traveled for more than a century. That is, the whooping crane recovery has advanced from conservation—keeping the last scraps of a species in place—to restoration, actually expanding the animal's presence in the world.

Operation Migration uses ultralight planes to teach the cranes to migrate along that new route. Traditionally, and for many millennia, whooping cranes learned to migrate from their parents; the route was passed from generation to generation. Unlike other birds, whooping cranes aren't colonial—a so-called flock doesn't fly south all at once, and juveniles aren't inclined to follow, or even associate with, older birds that aren't their parents. Instead, cranes travel with their mates or in nuclear families. In the fall, a male and female will guide only the chick they fledged the previous spring on its first migration. That one trip is enough to teach the young bird the way.

But the cranes being released in Wisconsin have been carefully bred from animals living in captivity. They hatch at the Patuxent Wildlife Research Center, a government lab in Maryland. In other words, they're plopped down in the wild without parents—they have no guides. The trikes act as a surrogate parent, flying ahead of a group of cranes, which have been trained, since birth, to follow them. As in nature, with real crane parents, it takes only one journey south behind an airplane for the birds to learn the way. After that, they will migrate back and forth independently for the rest of their lives and eventually teach their own offspring. Meanwhile, the ultralights move on to lead the next class of government-issued chicks—incrementally, and artificially, building up a population that acts like a wild one.

However, because the goal is to produce birds that, once released, will retain the majesty and uncompromising wildness that makes the whooping crane the whooping crane, those engineering this new population had to find a way to step in as surrogate parents and train the birds without wearing away the cranes' apprehension of people. Working with them too intimately would produce whooping cranes that feel comfortable hanging out in schoolyards, or in the fetid little ponds around retirement communities; it would plague America with a race of whooping cranes that are hardly wild or majestic but are more like the geese loitering on suburban soccer fields—except that, as Joe points out, a goose doesn't jump into the air and rake forward with its long talons when it feels cornered.

And so, to keep the birds wild, every effort is made, from the time the chicks hatch in Maryland and for the rest of their lives, to prevent them from encountering even the slightest sign of humans. No one who works with the cranes is ever allowed to say a word around them. And everyone wears identical white costumes: a long, frumpy gown and white hooded helmet with mesh covering the face. (All together, it resembles a bee-keeping suit crossed with a Ku Klux Klan getup.) They also wear a crane-head puppet on one hand, which they use to show the chicks how to peck and forage. The crane bonds to the white costume, just as a chick would to its parents in the wild—trusting and readily following it around. Scientists call the process "imprinting."

After a couple of months, the cranes are shipped from Maryland to Wisconsin, where Joe and Operation Migration's other two trike pilots—all wearing the same white whooping crane costumes, working the same puppets, and abiding by the same monastic vow of silence around the birds—put them through a summer of flight training. Back in Maryland, the newborn chicks were played recordings of an ultralight engine, sometimes while still inside their eggs, and attuned to it as a pseudo-parental presence. Now, hearing that sound again, and drawn to the costumed pilot in the cockpit, they eagerly scuttle behind a trike as it taxis on the ground. Eventually, they scramble into the air and learn to form a line behind the wing. The birds take longer and longer practice flights, spreading behind the airplane in a lopsided or often one-sided V. Then, in the fall, everyone sets out for Florida: the trikes and the cranes, a Cessna flying above them as a spotter, and an entourage of motor homes and support vehicles spooling along the highways below.

The caravan hopscotches from one pre-scouted stopover point to the next, keeping up the same wordless, costumed charade around the cranes the entire way. The birds are transferred clandestinely, priceless works of art. From the time they land at each stopover to the time they can take off again, they're stashed in a portable camouflaged pen, tucked out of view, and ringed by an electric fence. Costumed workers feed them twice a day, but otherwise the area is off-limits and resolutely shielded from trespassers, hunters, whooping crane lovers, journalists, and dogs.

I set out to follow one autumn's migration. And it was only when I got on the ground that I realized that the journey was even less straightforward than it sounds. This work—this wholesale manufacture of wild birds by human beings—turns out to be so ambitious, tedious, and packed with perplexing arcana that, after ten years, it's hard for those who have given their lives to the project even to agree on how well it's working and what they should do next. Under the surface, I'd find a simmering clash of theory and practice—and also a clash of personalities: the artists and scientists, vagabonds and bureaucrats that had thrown themselves together into the sort of improbable partnership that is capable of conceiving and pulling off something so bizarre, if it wasn't first swallowed by its own eccentricities. As in the time of Robert Porter Allen and George Douglass, the fate of the whooping crane hinged on the cooperation of very dissimilar people, the volatility of their individual spirits, and the resilience of their wills. At some point that fall, I would lose interest in the legendary wildness of whooping cranes and wind up wonderstruck, instead, by the wildness of human beings—whatever unfathomable force inside us that had once nearly destroyed the bird and was now beating just as uncontrollably in a cockpit at the front of the flock.

The journey from Wisconsin to Florida is 1,285 miles long. Many times, Operation Migration has left in the middle of October and not reached the end until late January. It's a sluggish, sometimes maddening way to travel the length of America—so baldly inefficient, so crazy-seeming on its face that, I'd notice, it often needs to be explained multiple times to the greasy-spoon cashiers and gas station attendants that Operation Migration encounters along the way. It's hard to believe. Already it's come to this on Planet Earth: men dressed like birds, teaching birds to fly.

THE SCREEN DOOR of Joe Duff's RV battered shut behind him. It was before dawn one morning in early October, at the migration's starting point, the Necedah National Wildlife Refuge, northwest of Madison, Wisconsin. "We're not going today," Joe said, and brushed past one of the Japanese guys who'd been waiting at the door for him to appear.

Joe is tall with silver-threaded, wavy hair, bright gray-blue eyes, and high cheekbones that give him a flash of boyishness when he chooses to smile. When he doesn't, like now, he can unintentionally project a detached and slightly smoldering intensity.

The three Japanese guys were television reporters. They were waiting to film the start of that fall's migration for what they described as Japan's equivalent of *Good Morning America.* Joe was away from the refuge when they'd arrived, and because reporters can get pushy and try to talk their way into putting on a white costume and getting close to the cranes, he'd warned one of Operation Migration's other pilots before he left, "Don't let them push you around." This turned out to be unnecessary advice. The Japanese men seldom spoke and dispensed a small wrapped gift to every person they interviewed. Just before Joe stepped out of his trailer, the smallest of the men had been milling outside the door, considering and reconsidering whether it would be too presumptuous to knock.

The Japanese guys had been standing by in Necedah for a week already. When they arrived, Operation Migration was still going through its final preparations: dismantling equipment and sorting through the kitchens of their RVs, separating the boxes of pasta that mice had crapped in over the summer from the clean pasta. But after a few days, everyone was ready to go, and it was only a matter of catching the right weather. Now Joe made it clear that the weather still wasn't right, and we'd all be waiting at least one more day.

The weather lords over Operation Migration like the gods in a Homeric epic, limiting or enabling their actions. Not every day that looks sunny and clear is a good flying day; not even most days are. Though wild cranes can fly the entire migration in as little as a week, swirling up to eight or ten thousand feet on columns of rising, warm air and coasting for miles, the trikes can't keep up with the birds at such altitudes and must stay closer to the ground, forcing the birds to flap the whole way. And both the trikes and the cranes have fussy, sometimes opposing requirements as to the kinds of weather and wind they'll fly in. Conditions need to be almost perfectly calm during a narrow window just after dawn. The slightest headwind can give the birds an excuse not to bother following the planes, or make the trip to the next stopover take so long that the birds won't have enough stamina to make it. Even a tailwind can be the wrong kind of tailwind: if the air is too choppy, the trike wings will wobble and jerk, and the birds, unable to lock in, will eventually lose patience and land.

Consequently, everyone in Operation Migration obsessively consults his or her favorite esoteric weather report, but ultimately, no one places trust in any of them. And so a routine was now solidifying at Necedah. Before sunrise every morning, OM's three pilots would stand outside at their campsite, under a sliver of moon, and stare silently at the trees, trying to discern any movement in the leaves, to gauge the wind speed and direction. Then the Japanese guys would pull into camp and hop out of their truck. One of them would say,

"Joe, how is the weather?" and because the weather was inevitably not good enough to fly, they'd all be back in their truck within minutes, heading for their motel room to confront another day without a single commitment in central Wisconsin. This went on for several days. It got a little baffling. One morning, for example, the pilots reached an immediate and solid consensus that the day was a no-go, based on the agitated rustling of the trees. But I couldn't see any rustling. "You can hear it," a pilot named Brooke Pennypacker whispered to me. I couldn't hear any rustling, either.

That fall, OM would be making its tenth whooping crane migration. There were, by that time, about a hundred older cranes flying around Necedah, the majority of them graduates of previous years' journeys. Having each been shown the way once by the trikes, the cranes had been migrating back and forth by themselves every autumn and spring ever since, set in motion like so many sweeping pendulums. They'd soon begin heading south again. Occasionally, you'd spot their waiflike white shapes standing in the reeds far away. It was impressive—all that physical, free-ranging proof of the reintroduction's success. And yet, as I was starting to understand, after a decade of work the entire project had now reached a dramatic, and not especially friendly, crossroads.

The reintroduction is overseen by a coalition of eighteen government agencies and nonprofits, including Operation Migration, called the Whooping Crane Eastern Partnership, or WCEP (pronounced "*Wee*-sep"). When WCEP formed, it was trying to help deliver the species from the throes of a crisis; the airplane idea, as odd as it looked, got more whooping cranes out onto the land, so they simply kept at it. "The reality," one partner at the U.S. Fish and Wildlife Service told me, "is that, for much of the life of the project, we've more or less done this by the seat of our pants." But despite having put some real distance between the bird and extinction by now, they'd never quite evolved out of that finger-in-the-dike mind-set. Earlier that year, an external review of WCEP had slammed the partnership for its "inertia to change . . . communication problems and internal politics." There was too much gut instinct, not enough empiricism. "Underlying many of these problems," the auditors wrote, "is a misunderstanding and poor appreciation of just what *is* science."

So WCEP was now taking a step back—to question some of the assumptions that they hadn't stopped to question before, run lots of studies, and think through the most efficient and cost-effective ways to go forward. At a series of meetings just before I arrived in Necedah, WCEP also totally reorganized itself, revamping the some-times aggravating looseness of the organization into a proper bureau-cracy, with different "teams," and flowcharts demonstrating how the teams were supposed to communicate with one another and make decisions.

All of this reassured the scientists in the partnership; they tended to like flowcharts and wanted to learn everything they could about whooping cranes and keep rigorously refining the process. But there are no scientists in Operation Migration. They're all laymen, all of whom had taken a circuitous route to crane conservation and up-rooted their lives over the last decade. They basically wanted to keep plowing ahead and worried that—ten years and a hundred whooping cranes into the project—the more scientifically conservative attitude rising within WCEP could unnecessarily undermine the momentum they'd built. "We are not scientists," Brooke told me. "We're con-struction workers. We're building a flock of whooping cranes. I'm not ashamed of that." The partnership had always been full of hard-headed, passionate people—idealists with differing convictions about how to do the most good. (One member recalled facilitating a two-hour debate about whether to outfit the cranes with tracking bands that snapped around their legs, or bands that were glued on.)

But negotiating this turning point had made a long-standing under-current of competitiveness between some of the WCEP partners even more glaring. There was especially a fairly widespread resentment of Operation Migration.

"OM has the ability to piss a lot of people off," one government scientist in the partnership told me. Joe Duff, by his own admission, wasn't always the easiest guy to work with. He knew he could be inflexible and in the past had been quick to lose his temper. But he also felt his team had an unfair reputation as glory hounds and ego-maniacs. (They always got the majority of the media attention, sim-ply because they were the ones flying the airplanes in front of the birds, he said.) Meanwhile, three key people at the Fish and Wild-life Service who'd been Operation Migration's biggest champions in the partnership were now retiring. Joe had taken to calling these men "the patron saints of whooping cranes." They'd understood that little about what was being attempted, and about some of the people at-tempting it, fit neatly into the strictures of the existing wildlife bu-reaucracy. But for them, saving the crane was deeply personal—the central work of their careers. And so, without cutting corners exactly, they'd always found a way to move the recovery forward and keep the peace. Their replacements weren't necessarily so accommodating or forgiving. One partner described the new guard to me as "dedicated bureaucrats," rather than dedicated conservationists. "They want to preserve their jobs."

One of these newcomers was Doug Staller. Staller, a high-energy guy with a military crew cut, had recently been made the new man-ager of the Necedah National Wildlife Refuge by the Fish and Wild-life Service. He was under the impression that his predecessor had cut Operation Migration too much slack. And so, that summer, Staller started restoring order. He had, for example, barred OM from train-ing the cranes in certain areas of the refuge, claiming that the ultralights were bothering other species of birds. I'd noticed whooping crane banners on the light posts up and down Necedah's struggling main drag. The wireless password at my motel was "whooping-crane." But Staller didn't seem to care about that sort of charisma or feel a special obligation to the whooping crane; instead, he kept invoking the "congressional mandate" of the refuge to be a sanctuary for all migratory waterfowl, equally. When we met, Staller even questioned why WCEP was going through all the fuss, dressing up in costumes and such, to pamper what were supposedly wild animals. "Actually," he told me, "I have thought long and hard about bringing my two big Labradors out here on a long leash and—not catching a crane—but just chasing them," to toughen them up.

I'd gone to see Staller my first morning in Necedah because, whereas guests used to come and go freely from Operation Migration's little encampment of RVs on the refuge, Staller now required everyone to have a permit, issued through his office. He seldom issued them. He told me he worried that journalists like me would hang around with OM and get "seduced" by their celebrity, taking their side in the disagreements now weighing down the partnership. "This is biopolitics at its worst," Staller said.

Staller was noncommittal when I asked him for a permit. Instead, he took me on a short drive around the refuge. We stopped periodically to look at ducks—it was obvious that Staller really likes ducks—and checked out some ditches and dikes.*

---

* Necedah's wetlands were drained for agriculture a century ago. But, as a Fish and Wildlife Service history of the refuge puts it, "after a series of intense peat bog fires in the 1930s, many settlers abandoned their homesteads." So, in 1939, the land was flooded again as a public works project and turned into a wildlife refuge. Its wetlands still need to be meticulously flooded and drawn down to maintain just the right habitat for birds. You'd never know by looking at it, but the refuge is an artificial landscape—a man-made terrarium with no walls or roof.

I spent the rest of the day running errands with Operation Migration. But I came back to Staller's office before quitting time. Staller, knowing I'd been with the crew, stopped me in the doorway and looked into my eyes for a long, uncomfortable moment. Finally, he said, "You don't *look* brainwashed." Then he cracked a conspiratorial smile and handed me my permit, which is why I was able to stand at Operation Migration's camp every morning and stare at the treetops in the dark.

THE FIRST DAY OF MIGRATION, when it finally came, started like every other day except that, after peering into the leaves for twenty minutes, the pilots decided that the weather looked good enough to give it a try, and everyone filed into vans and pickups, carrying their white whooping crane costumes under one arm, heading for the ultralight hangar.

Brooke was taking the first turn as lead pilot. While the other two trike pilots hung high in the air as spotters, he circled down toward an area of the refuge called Laskey Field, where the cranes had been shuttled a few days earlier as a staging area for departure. He landed so that his propeller faced the door of their pen and started broadcasting an MP3 of a whooping crane call from a bullhorn on his rear axle— the same croaking call that had been played to them at Patuxent. Then, with the birds popping up and down and tensing their wings inside the pen, two costumed interns opened the gate and rushed out of the way.

There were ten whooping cranes, all about six months old and four feet tall, still with patches of cinnamon-colored juvenile feathers that had yet to give way to white. They gathered in a mob behind Brooke's trike as he rose out of Laskey Field. He managed to get four clinging to the tip of one wing, but within seconds they fell behind and scattered with the others. A few birds found their way back to Brooke, but again and again new cliques of rebels—three or four cranes at a time—kept breaking away.

They were heading for Site 4, the area of the refuge where they'd been penned for most of the summer—they were, in other words, going back home. Brooke kept circling back and maneuvering the edge of his wing in front of the cranes, trying to lure them. But any progress he made never lasted. The birds flapped away again. The formation would suddenly slacken, then dissolve, the way a school of fish dematerializes. "I'll come down on the outside of them," I heard Joe say on the radio at one point, swooping in to cut off a few deserters.

It was chaos. It was getting difficult even to keep track of the cranes. There were a few older whooping cranes in the vicinity, too, and also lines of wild sandhill cranes—the other species of North American crane, slightly smaller than the whoopers—commuting through the trikes' airspace. Eventually, Brooke simply decided to follow the birds back to Site 4 and land with them, giving them a rest. He would shut down his trike, kneel on the ground in his costume, and hand out a few grapes for a snack; he'd let *the birds* feel in control for a moment, then try to get airborne again. But as soon as he landed, he saw one of the cranes hobbling at the other end of the grassy runway near the door to the old pen. Brooke was worried: the legs of young whooping cranes are fragile.

By now, Joe reported, the wind had shifted. The tailwind had become a headwind—a strong one. The first migration stop, a small family farm, was twenty-three miles to the south, and it was clear that they'd never be able to push that far even if the cranes suddenly cooperated. It would be another down day after all.

So Joe landed at Site 4, too, got out of his trike, and walked right past Brooke toward the injured bird. The scaly exterior at the bottom of its leg was torn off and raw. The bird needed to be coaxed into a crate and driven to a vet—it probably needed an X-ray. They whispered into the radio for another team member to come help and started herding the rest of the cranes into their old pen. They'd pen the birds here now and try to depart again tomorrow morning—weather permitting, of course.

Meanwhile, there was trouble back at Laskey Field, too. At the outset of the flight, Joe had spotted one crane that never got airborne with Brooke and asked OM's third pilot, a Canadian metal sculptor named Richard van Heuvelen, to circle back and cajole it into the air. "I don't know who it is," Joe radioed, "but it never did fly." It turned out to be crane number 2, the biggest and most willful of the cohort. Richard could see it, still lingering at the door to the pen. Richard considered his options. Then he called for someone on Operation Migration's ground crew to let out the Swamp Monster.

If a bird refuses to launch, or abandons the lead trike and lands back at the pen, an OM operative will often throw a brownish tarp over him- or herself and run at the crane, crumpling the tarp loudly or blasting an air horn to scare it back into the air. They call the maneuver the Swamp Monster. But this time, when one of OM's interns Swamp Monstered crane number 2, the bird didn't shoot skyward and latch onto Richard's wing; instead, it made a beeline straight into the oak trees bordering Laskey Field and hid. The intern took off the Swamp Monster tarp and pursued the crane on foot in his regular white costume—he was dressed as the reassuring parent figure again. But number 2 kept retreating farther into the woods. The intern kept approaching. Number 2 wanted no part of him. It was as though this bird had had a terrible epiphany.

This could have been the first day of migration. Instead, one crane was now injured and a second had apparently experienced some sort of crippling psychotic break in midair. And rather than migrate twenty-three miles to the first stopover point, Operation Migration managed only to take the birds back to the same pen they'd transferred them out of a few days earlier. "We went backwards," Joe said.

12.

CRANIACS

n the early 1980s, thirty-five years after Josephine's trysts with Pete and Crip at the zoo in New Orleans, America found itself swept up in another high-profile whooping crane love story—another bit of desperate matchmaking, which, if it worked, might afford the species a healthier future. The circumstances of this new couple, Tex and George, were even stranger: Tex was a female whooping crane; George was a human being.

At the time, the government biologists at the Patuxent Wildlife Research Center were honing their skills at breeding whoopers in captivity, trying to extract as many genetically healthy new birds as they could from the shrunken gene pool. Tex, who'd hatched at the San Antonio Zoo in 1967, was valuable in this regard. As the sole offspring of two captive cranes that had lived in the wild, she potentially carried genes that no other captive bird did. There was a problem, however. Tex's sibling had died after hatching—one account describes Tex's mother sitting on the chick. So, as a precaution, the zoo's director snatched Tex into protective custody, raising her for six weeks in a cardboard box in his family's living room. Having barely seen her parents, or any other whooping cranes, Tex was left to imprint on the one animal she did see: the zookeeper—a dark-haired white man of medium build named Fred Stark. ("This is what happens to a lot of animals," one researcher told me, explaining how easily imprinting can go wrong. "You know 'Mary Had a Little Lamb'? It followed her everywhere? Well, the lamb imprinted on Mary.") In short, Tex believed she was a human. She wouldn't mate with other captive cranes—she wasn't interested. But when a dark-haired white man of medium build walked past her pen, she would holler and grind.

George Archibald happened to be a dark-haired white man of medium build. A young ornithologist who'd worked with whooping cranes at the Patuxent Wildlife Research Center in Maryland, he had recently founded a nonprofit, the International Crane Foundation, with a friend in Baraboo, Wisconsin. In 1976, he convinced the government to ship Tex there, and when mating season started in the spring, George subdivided Tex's enclosure with chicken wire and claimed half of it for himself. He put in a cot and lived there as Tex's companion. They passed entire days together, going on walks. And they danced. George would do deep knee bends and spring up with his arms out like wings. He'd issue loud whoops and shout, "Come on, Tex! Come on!" And soon he and Tex would be dancing together, just as wild cranes do during courtship. This would get Tex aroused—*People* magazine described George's dancing as bringing Tex to "a fervid emotional climax"—and then when the crane stood motionless and extended her wings, two of George's assistants would rush out from a hiding place with a syringe and inseminate Tex with the semen of another captive crane, flown in fresh twice a week from Patuxent.

George did this for three consecutive springs, because each year either the egg that Tex laid was infertile or the chick died while hatching. In 1982, he went back at it—for six straight weeks this time. He didn't enjoy it. He was miserable, actually. ("I was getting a little bit weird," he told me when I met him in Baraboo, though he wouldn't elaborate.) He had a small shed at his disposal, to stay out of the rain, with a table, a typewriter, and a stack of books. But he spent a lot of time just staring numbly at the grass. That year, however, Tex laid an egg with George sitting beside her, and in early June the egg hatched. George named the chick Gee Whiz. A *Los Angeles Times* headline read: "Man, Crane Proud 'Parents' of Chick."

George was flown to Los Angeles to appear on *The Tonight Show* with Johnny Carson. He knew the publicity was valuable, but suspected that he'd been invited on the show only to be made fun of. He is a private person, he told me, and understood how strange, or even indecent, the affair looked. "It was embarrassing for me. I mean, here was this guy living with a crane." One of his heroes, the visionary ecologist Aldo Leopold, had written a famous essay about cranes in which he calls the bird "wildness incarnate"—a resident of an "untamable past" that we humans left behind. But here, one of those stately birds had been so pitifully marooned in the modern human world that it was practically married to a man—it was about to become a punch line on late-night television. While George waited in his hotel room in Los Angeles, leafing through his Bible for some reassuring scripture, the phone rang. It was a colleague back in Baraboo, calling to tell him that Tex was dead: a thirty-five-pound raccoon had torn into her pen and ripped her apart during the night.

As it turns out, the tragedy spared George any humiliation on television. It elevated the saga of George and Tex beyond ridicule. After fielding some questions on the air that night, George composed himself and announced, "Last night, while I was on my way to California, Tex was killed by a predator." The studio audience was stunned and stopped giggling. Twenty-two million people were watching, and the poignancy of that moment arguably brought whooping cranes to a level of visibility and sympathy that they hadn't had since the days of Josephine, two generations earlier. The story of George and Tex, a *Washington Post* editorial noted, showed what hard work could accomplish at a time when species were vanishing at a rate of one per day. "Gee Whiz's birth," the paper wrote, "is one tiny step in the other direction."

But the truth was that no one working on the whooping crane recovery knew what the next step would be: how they'd ever manage to get these psychologically bungled whooping cranes out of the lab and into the wild, released from human custody untainted and unattached. It was a peculiar puzzle to have to solve. And as you might imagine, their ultimate solution—hiring airplane pilots in white suits—was reached at the end of a long chain of other, only marginally less unnerving ideas, each one building on the last.

OPERATION MIGRATION traces its roots to a summer morning in 1985, three years after George's appearance on *The Tonight Show,* when a man named William Lishman took his ultralight for a ride.

Lishman lives in a rural area of Ontario, northeast of Toronto, in a house that he designed himself. It's a circular, domed cavern, almost completely underground, which, Lishman has said, provides his family "the security of a cave and the coziness of a cottage." Lishman is an artist. His sculptures include a full-scale replica of Stonehenge built with old cars, and a life-sized metal Arnold Schwarzenegger, balled up as a fetus inside a twenty-two-foot-tall wire-frame uterus—a piece Lishman calls *Hasta la Vista Baby.*

Lishman had wanted to fly ever since he was a boy. All humans fantasize about flight, he says—it's natural to look up at birds and feel envious. But Lishman's feelings about birds move beyond envy, almost into reverence. In a memoir, he goes so far as to speculate that maybe, millennia ago, birds evolved into a technological civilization like our own, but then "thought it out well enough" and foresaw the kinds of societal and environmental problems that such a civilization would inevitably create for itself. So, instead, the birds learned to "engineer their own genetics" and evolved—albeit, seemingly backward—into the creatures we see today. They traded modernity for boundless, liberated lives. "They have no alarm clock making them report for work each day," Lishman writes. No income tax or laundry. "Their comfort is self-contained, the only upkeep their feathers," and they swish across the earth, following the best weather. He imagines all birds linked together by a vaguely telepathic "bird Internet," and a flock as basically being a V-shaped kibbutz, where each individual can "enjoy the camaraderie of their kind and fly with far greater efficiency than if they traveled alone."

I mention all this background to give some inkling of the mind-bending euphoria and freedom that Lishman felt that morning in 1985 when, flying his ultralight, he accidentally startled a flock of ducks into the air and found himself momentarily flying among them. The wild-looking man with a bristly brown beard was suddenly soaring inside a cloud of birds, sharing the air with them. It was as though he'd forged a living sculpture. He craved that closeness again. So he learned about imprinting and raised a brood of Canada geese, imprinting them on himself so that they'd follow him and his plane. Soon he was leading the geese on regular flights around his property, confusing and thrilling his neighbors. In 1989, he produced a short video called *C'mon Geese*, which is what he yelled at the birds from his cockpit to get them to follow.

It had now been seven years since George Archibald's work with Tex, and those dreaming of reintroducing whooping cranes—of making them wild—had surmounted a few, but not all, of the roadblocks in their way. A big breakthrough had come after a psychologist named Robert Horwich approached George's International Crane Foundation believing that a whooping crane's freakish willingness to imprint on virtually anything might be turned into a tool to help the species, rather than just a hazard to be worked around, as it had been with Tex.

Horwich had never worked with birds before; he studied early psychological development in humans and other primates. But he suspected that young cranes went through the same developmental patterns of attachment as primates, and that understanding those phases would allow him to hack into the birds' psychology, and manipulate them, by posing as their parent. To keep birds from imprinting on their human caretakers, researchers at the International Crane Foundation had started feeding newborn chicks with bird-shaped puppets, poked through the holes of high-walled boxes. Now Horwich wanted to actively—*intentionally*—get cranes imprinted on him. But he would wear a costume and never speak. This would allow him to act as the cranes' surrogate parent for longer, teaching them more and more as they grew. The birds would never know that the creature they were following around was a human. George Archibald told Horwich to try it—he was not in a position to laugh at odd ideas. But George added, "I don't think cranes can be that dumb."

In 1985, Horwich was given ten sandhill cranes, as proxies for whooping cranes, with which to run his experiment. He led them around the marshes at Necedah, wearing a gray sack and mask with some feathers sewn on the wing-arms. Soon he taught them to fly, running alongside them like a boy launching a kite. That fall, nearly all of Horwich's sandhills followed older sandhills on migration; they returned to the refuge the next spring. They were still afraid of humans—still wild. Horwich hadn't left a mark.

Still, conservationists were aiming to establish a brand-new population of whooping cranes, geographically separate from the last surviving one out west. And, unlike with Horwich's sandhills at Necedah, in such a reintroduction there wouldn't be any older whooping cranes around for the first, costume-reared generations to follow on migration—no older, more knowledgeable animals to show them the way. And even if there were older whooping cranes around, it seemed unlikely that they—not being a colonial species like sandhills, and bonded to only their own chicks—would tolerate these young interlopers, or that the younger birds would be interested in following the older ones in the first place. (Even now at Necedah, older whooping cranes often harass or attack Operation Migration's latest class of juveniles.)

For years, there appeared to be no way around this problem. (Scientists had already tried outsourcing the job to sandhill cranes, switching out the eggs in sandhill nests in Idaho with whooping crane eggs and hoping that the more cordial sandhills would adopt the whooper chicks and teach them to migrate. But the young whooping cranes turned out to be sexually confused by their surrogate parents, in the same way that Tex had been by her zookeeper: they were attracted to the wrong species. No new whooping cranes ever hatched during that fifteen-year experiment, though the birds did manage to produce one "whoop-hill" hybrid.) Then, one day, long after Robert Horwich had moved on from his job at the International Crane Foundation, he saw a picture in a magazine of an ultralight plane leading a flock of geese. The caption under the photo said very little about the ultralight pilot. But Horwich, grasping at another far-fetched idea, tried to write the man a letter anyway. He addressed the envelope: WILLIAM LISHMAN, THE GUY WHO FLIES WITH GEESE, BLACKSTOCK, ONTARIO CANADA.

———————

By that time, George Archibald had been shown Lishman's video, *C'mon Geese,* by a friend and was arranging meetings between the artist and other leading crane conservationists. Lishman had only wanted to fly with birds because it made him feel amazing, but he was intrigued that his new skill-set might be repurposed to do some good. It was unclear whether an ultralight could actually teach birds a migratory route, so, after some false starts, it was decided that, in the fall of 1993, as a test, Lishman would lead eighteen geese from his property in Ontario to a bird sanctuary four hundred miles away, in Virginia. Lishman needed another pilot. He asked a buddy of his from the ultralight circuit—a thrill-seeking, slightly cocky car photographer with a star in his tooth named Joe Duff.

Joe and Lishman made it to Virginia in seven days. The following spring, thirteen of the geese returned to Ontario on their own. The scheme worked, in other words: the birds had learned the route and could make the return trip themselves. The two pilots set up a nonprofit, Operation Migration, so they could fund more experiments and, over the next several years, did a number of progressively more ambitious migrations with geese and sandhill cranes. They migrated farther and farther, with more and more birds, then gradually phased out all talking during their trips and incorporated Robert Horwich's costume idea to keep the birds from getting tamed down in the process. One autumn at a time, Duff and Lishman were building a case that—no matter how it might have looked to the Canadian and U.S. government agencies responsible for the whooping crane—these two artists in their funny airplanes could help resurrect one of the world's most endangered birds.

The 1990s was actually a period of fervent experimentation in North American whooping cranedom, with government scientists and graduate students grasping at only slightly less far-fetched ideas to overcome the migration obstacle. They tried driving sandhill cranes—again, as a stand-in for whooping cranes—on a migration in a horse trailer. When the birds didn't fly back, they drove them again, but stopped every fifteen or twenty-five miles along the way to let the cranes fly around for a few minutes, hoping this would orient them enough to connect the dots later. There were multiple attempts to imprint sandhills on an ambulance and then have the birds fly a migration behind it, swooping over back roads like police helicopters in an action movie car chase. Unfortunately, on one such trip, many birds got injured or killed when they flew into the power lines beside the road. (The researchers had mapped out a route that forced the birds to cross over power lines hundreds of times.)

By the end of the decade, there had been more than twenty different attempts to lead birds on a migration behind some sort of motorized vehicle. They tried everything. (At one point, a Russian artist was even commissioned to build a remote-controlled, super-sized whooping crane robot that could fly at the head of a flock.) But they ultimately conceded that nothing worked as well as the ultralights. And in the fall of 2001, a joint panel of U.S. and Canadian government agencies entrusted Operation Migration with whooping cranes, giving Joe and Lishman ten birds for their first migration to Florida. OM made the trip in forty-eight days. Joe told me that it's still the fastest they've ever done it.

Joe was narrating this history for me over dinner one night at a steak-and-seafood joint in Necedah called the Reel Inn. The truth was, he explained, he'd fallen into conservation inadvertently. When Lishman asked him to join the first goose migration, back in 1993, he was newly divorced and bored. He'd started to suspect that he'd taken the life of a luxury car photographer as far as it could go, and was considering selling his studio and starting a marina in Ontario's lake country with a friend. Joe had been watching Lishman work with the geese and was seduced for the same, essentially selfish reasons as his friend—for the thrill of flying with birds. But at some point, Joe looked up and found himself attending meetings of government biologists. He was filling out bureaucratic permit applications to wildlife agencies and typing up scientific papers about the ultralight migrations and publishing them in *Proceedings of the North American Crane Workshop*. He was committing himself to the work, in other words, with a determination that not many people who knew him at the time would have expected. Lishman would eventually get tired of the migrating lifestyle; he stopped flying with Operation Migration in 2004. But Joe was too invested to stop, even as he remarried and had a child, rebuilding his personal life in ways that made him want to stay closer to home. When I met Joe in Necedah, his daughter had just turned eleven. "It's kind of normal for her, my being gone," he said. "It's getting harder, though." Almost every autumn of her life, he has told her goodbye and spent the rest of the year flying away from her—very, very slowly.

OPERATION MIGRATION FINALLY got its eleven whooping cranes out of Necedah two days after that botched first try. The trip through the rest of Wisconsin was typical and slow. There were fits of flying broken up by wind and rain, which stranded them at stopover sites for days at a time. On day nine, they crossed the border into Illinois, the first of many modest milestones on migration. But almost as soon as they did, the Midwest was blasted with a weather system of biblical fury, and OM found itself grounded northwest of Chicago, in Winnebago County, for what looked to be a long, long time.

There are chores to do on down days: feeding and checking the cranes, flushing the bowels of the RVs, and keeping a supply of fresh pumpkins in the pen for the birds to spear and stomp to stave off their boredom, and to keep them from spearing and stomping each other.* Crew members and other WCEP partners also visit schools along the route, fielding enthusiastic questions from little kids while standing in front of the blackboard in their full white regalia. (Classes around the United States and Canada follow each year's migration on the Internet as part of their science curriculum.) When I tagged along on one visit, to a third-grade classroom in Reedsburg, Wisconsin, an Operation Migration intern closed her presentation by inviting the children to feed each other grapes with the beak of her crane puppet.

In the end, the crew wound up grounded in Winnebago County for eleven days, a new Operation Migration record for immobility. By the time they made the next stop, the migration was in its twenty-first day overall, and they'd traveled a total of 185 miles from Necedah— about a four-hour drive.

But OM slowly gained momentum through Illinois, soaring over the endless, checkered grid of early-harvested fields and the monstrous wind farms that, over the years, have multiplied in their path. Soon they were crossing into Kentucky, and south through the state into Marshall County, just shy of the Tennessee border. There they set the birds down on a soy farm owned by a boisterous old widow named Martha.

The landowners at every stopover have volunteered their property and get no compensation for letting the birds and the half-dozen motor homes and trucks accompanying them invade their land. They

---

* Whooping cranes are not particularly nice. The politics inside the pen are like those inside prison, with dominant birds making their reputations by bullying ones that show weakness. One year, Operation Migration hired a short Eastern-European woman with an unassuming posture as an intern, and whenever she suited up and got in the pen, Joe told me, "the birds used to whale on her." Once, after a whooping crane died in the pen, the other cranes pecked out its eyes.

don't know when OM will arrive exactly, and have no idea how long the weather will force them to stay. And, like everyone else, the landowners are prohibited from even seeing the birds and wind up barred from the portion of their own property where OM has hidden the pen. On top of everything, they're asked to keep the entire affair a secret—no bragging to friends and neighbors, or posting pictures on Facebook—since, for security reasons, the public is not supposed to know the cranes' exact where abouts. (OM asked that I not use any of the landowners' full names.) Still, over the years, the crew has grown close to many of their host families, cobbling together genuinely warm relationships from just the few days when they visit together every autumn. For the crew, it can feel like stepping into a family portrait—a living Christmas card—and noting small differences from the previous year's snapshot: Kids get bigger. Friendly old dogs disappear. A few years ago, OM was rerouting the southern half of its migration to avoid some tricky terrain and had to scout and recruit new stopover hosts. One of the pilots, Brooke Pennypacker, drove around knocking on doors with his girlfriend, explaining the whole story—the costumes, the airplanes, the top-secret flock of imperiled birds—trying to sound as sane as he could. And it was amazing, Brooke told me, how many people welcomed him in right away and said yes. Few of the hosts are hard-core environmentalists. "But they're like a lot of people: they would help if they had the opportunity."

That kind of generosity fuels Operation Migration's journey. The group gets little, if any, government money every year and has to scrape most of its budget from individual donors. A fan base of so-called Craniacs now follows the reality show–like drama of the migration every fall on OM's Web site. A small camera mounted on the lead trike streams live footage to the Web, and Operation Migration's blog pumps out updates several times a day. There are dramatic recaps of flights, stories about the antics of individual birds in the cohort, and news about other, older whooping cranes in the flyway who've graduated from the ultralight program and are now on their own—sort of like the wedding announcements and updates in the back of alumni magazines.

In short, OM has found creative ways of making ordinary people feel close to a species it is also, simultaneously, devoted to keeping as far away from people as possible. They've created an intimacy that's entirely vicarious. The only time the public is even allowed to catch sight of the birds is during early morning "flyovers," when, setting out for the next stop, the pilots lead the cranes over a predetermined gathering point, like a parking lot or wildlife sanctuary. Folks assemble at the flyover locations before dawn, bundled in their parkas, wiping the frozen sleep from their young children's eyes. Many have read enough online about the individual cranes' personalities to gossip knowingly about that year's class. They buy T-shirts and DVDs from Operation Migration's director of communications, Liz Condie, and they hand over care packages for the crew. The gifts are almost always food: heavy hams, pies, pots of chili. "Everybody puts on ten pounds on migration," Liz told me one morning, after, in the span of two or three hours, I'd watched her collect several dozen donuts, three batches of homemade fudge, and a carrot cake. No one ever thinks to bring them a salad.

I met a lot of Craniacs at flyovers, and I liked them all. I met bird watchers and waterfowl hunters; veterinarians, tree farmers, receptionists, and stay-at-home moms. I met ex–military pilots and aviation buffs who like to geek out about the specs of the trikes. I met a woman who got enchanted by whooping cranes while recovering from brain surgery—"This is my life now," she told me, "see wildlife and learn"—and a man who worked for a washer-and-dryer manufacturer and had driven two hours to see the departure from Necedah in an SUV with one bumper sticker that said "We Love Whoopers" and another that read "Certified Craniac." And in southwestern Kentucky, outside a tiny white church, I met a bespectacled novelist named Squire Babcock who was lingering in the twenty-three-degree weather, still staring over the bare, black field to the spot where, a few minutes earlier, the trikes and birds had disappeared at the horizon. Babcock had been invited to the flyover by a friend and, not knowing much about the project, was still trying to get his head around what he'd seen—how utterly selfless it was, how completely *not* geared toward humans, at a time when nothing on Earth seems to be not geared toward humans. "It's such an intensive effort," he said. "And it's just about the birds. It's not about the people doing it." Babcock could hardly fathom it. "This is almost like a kind of spiritual statement on the part of these people who are doing this," he said. "These people are heroes."

The commanding, even mystical allure of the project was apparent as soon as Operation Migration made its first experimental trips south. The simple image of an aircraft leading birds riveted people— just the sight of it. Especially children. When they flew over upstate New York with that first cohort of geese, Lishman remembers, kids poured out of an elementary school to watch them pass overhead and school buses pulled over on the side of the road.*

The whooping crane has been an iconic species for generations, a symbol of all dwindling wild things and places. But now it's as though

---

* In 1996, Columbia Pictures produced *Fly Away Home,* a family film loosely based on William Lishman's memoir *Father Goose.* In it, a young girl, played by Anna Paquin, and her eccentric inventor father, played by Jeff Daniels, rescue a group of goslings and save a wetland from development by leading the geese on a migration behind an ultralight. (The plot is complicated.) The psychologist Gail Melson describes the film as part of a modern genre, like *Free Willy,* that shows children "as the true stewards of embattled nature," and "as allies and often saviors of vulnerable animals against an unfeeling, cruel adult world." In the film's most transcendent scenes, where the little girl flies with the geese, it is actually a friend of Joe Duff's named Jack Sanderson, in the cockpit, wearing a helmet with fake pigtails sticking out of it.

the ultralights and cranes had fused into a new form of hybrid charismatic megafauna—an indicator species, helping people situate themselves emotionally on a planet that feels as if it's losing its wildness completely. The old relationships have changed. As Joe put it, "The creatures that taught us the art of flying are now being helped by the aircraft they helped design." It's disorienting. But it's also inspiring, in a way that seeing a flock of wild whooping cranes soar by, alone, might not be.

I'd felt it myself when, just after sunrise one morning at Necedah, I first saw a trike ushering the cranes across a foggy marsh. It was the exhilarating weirdness that comes from suddenly understanding that more is possible than you thought. It might be like what watching the plains ripple with buffalo once felt like, or the sky darken with passenger pigeons. Maybe it was what the men and women of the Pleistocene felt, twelve thousand years ago, taking in the scale of North America's mammoths and automobile-sized armadillos, just before they started bringing those animals down with their spears. Philosophers talk about "the sublime"—about being diminished by a beauty larger than one's self. But this is a little different: it's the beauty of humans trying to fix a larger beauty we broke.

It never got old, the sight of an airplane leading birds. It can even feel weirdly familiar, like a picture from a bedtime story. Joe told me that, once, when his daughter was four, she turned to a friend and asked, "What kind of bird does your dad fly with?"

13.

THEIR INCREDIBLE ESSENCE

There's a very inconvenient postscript to the story of Humphrey the Humpback, the whale that got lost near Antioch Dunes. And I suppose I held it back because I worried it might spoil the uplifting resolution of the rescue's final scene: how crowds lined the Golden Gate Bridge in the fog to watch the whale return to the ocean; how Humphrey paused underneath them to swim in circles, leap repeatedly out of the water, and slap its tail as though, as Isla's picture book about Humphrey's adventure described it, the whale was "saying goodbye and thank you to all his friends who had helped save his life"; and how a government spokesman—relieved, though still stumped about why the whale came upstream in the first place—could only tell the media, "Now all we can do is keep our fingers crossed and hope he doesn't come back."

The problem is, Humphrey did come back. One morning five years later, in October 1990, the same humpback, identified by markings on its tail fin, swam into San Francisco Bay and promptly beached itself in the mud beside a low-lying stretch of highway near Candlestick Park, tying up rush-hour traffic. It took only two days to turn Humphrey around this time, though the whale nearly died in the process. People tended to the animal around the clock, draping wet towels over its exposed skin.

The incomprehensible persistence with which Humphrey had now twice, unmistakably, swum straight into civilization made it clear how little we knew about the whale. (For one thing, scientists only recently discovered that Humphrey was actually a female.) That void was filled with lots of wide-ranging speculation. One theory was that Humphrey was a forerunner of a new species, trying to take an evolutionary leap onto land. It was the not-knowing that made the animal's predicament even more painful to watch. "I just wish we could talk to him," one woman on the side of the road told the press.

I thought about Humphrey while tracking the long genesis of Operation Migration's work—how, over two decades, the entire collage of peculiar ideas had been pasted together into the costumed spectacle I was now watching. Layered inside that history, I picked up on a certain longing for closeness and collaboration with animals— for mutual understanding—and for proof that the wall between our world and theirs is an illusion; that we can enter their world, and help them, and solve problems by working closely together. It's what made George Archibald's dancing with Tex feel magical to the public, even as George himself felt beaten down by the work. And it's what Squire Babcock seemed to be moved by, but still struggling to express, outside the church in Kentucky. Ultimately, what all those Craniacs and I were gathering in the cold to watch was really a kind of interspecies communication. It was a very rudimentary and not so romantic kind, but something *was* being genuinely said and understood. The white costumes were saying, "Follow me."

Lately, a small genre of YouTube videos has centered on these momentary and almost metaphysical close encounters between people

and wild animals: The woman who manages to cuddle with an elephant seal on a beach, for example. Or the middle-aged American
tourist who sits stock-still on a rock in Uganda while a troop of mountain gorillas calmly encircles him. Soon the juvenile gorillas clamber
up the back of the man's shirt to inspect and groom his silver hair, as
though initiating him into their ranks. The astonished, euphoric look
on these people's faces is always the same. ("I'm a gorilla!" the tourist
says in a quavering whisper when the gorillas have finally trudged off.)
It's a look we'd never expect to see on someone rooting around with
pigs in a barnyard, or even nuzzling a pet golden retriever. That glee,
that flash of illumination, only seems to come from finding equal
footing with the wild ones. We believe that they have something to
teach us—that glimpsing what is common between us and the animals elevates, rather than degrades, our humanity. It's why William
Lishman first raced into the air to fly with geese in the summer of
1985, and why, a few months later, clear across the continent, so many
ordinary people felt called to help Humphrey the lost whale—and
why they felt called to help Humphrey again, five years after that. For
most of American history, whales were seen as commodities to be
cracked open for their oil. But by October 1990, even an average commuter from San Jose, standing on the side of the highway, could be
found making sense of Humphrey's return by telling a reporter from
USA Today, "Whales are so intelligent. I think they're connected
somehow to a supreme intellect in the universe, and he's here to help
us somehow."

I don't know when this reverence for animals was forged. Lots of
indigenous cultures, of course, felt roughly the same way about the
animals they encountered. But if there was a tipping point, a moment
when it was obvious that such feelings were permeating the mainstream, it was probably one afternoon in the mid-1970s when a woman
named Joan McIntyre was paid a phenomenal sum of money to speak

at a retreat for IBM's "Golden Circle" of top-performing salesmen in San Francisco. With trippy footage of whales and dolphins projected behind her, she looked out on what appeared to her to be two thousand men in identical suits and haircuts and began to explain, rather rhapsodically, how liberated these animals are, and how they make love in all possible combinations: mothers and sons, fathers and daughters, brothers and sisters.

It was the era of free love, and McIntyre—a conservationist cut from a new cloth—saw no reason why humanity couldn't extend its love to whales.

JOAN MCINTYRE, then in her early forties, was a glamorous and sharp-witted hippie from Berkeley who still carried the spirit of a twenty-two-year-old. She'd recently spearheaded an antifur campaign for the environmental group Friends of the Earth, and now had a large grant, no strings attached, to pursue a new cause.

Her interest in whales was piqued, in part, by *Songs of the Humpback Whale,* a record album of male humpbacks' haunting croaks and bellows released in 1970.* In an interview, McIntyre later explained that something about those moaning animals tapped into feelings of loneliness that she'd carried with her since childhood. Her father died when she was very young, and her mother shuttled her between Chicago and Los Angeles, chasing work during the Depression. She was a solitary child and disappeared into the many books about fictional

---

* The recordings, made in Bermuda and Hawaii by Roger Payne and Scott McVay, represented a step forward for science—few people, aside from top-secret navy engineers, knew that humpbacks vocalized. The American counterculture was also drawn to whale song as though it were nature's lava lamp—a swirling psychedelic bath of sound to lose oneself in. Even the music critic for the *New York Times* suggested listening to *Songs of the Humpback Whale* in a dark room and getting "in touch with your mammalian past."

animals that were popular at the time, nature faker–esque tales whose animal protagonists were often friendless or abandoned. When she got involved in antifur activism in the late sixties, as an adult, it was because she could not stop herself from feeling empathy for those fur seals and minks—actually *feeling* it. It was as though, at some point in her youth, McIntyre had learned to see the world of animal sentiment and suffering she'd read about in storybooks hidden in the actual world.

When she set out to combat the commercial whaling industry in the early seventies, most nations' whaling fleets had harpooned themselves out of a job; stocks of many species were driven so low that hunting whales was unprofitable. But the Russian and Japanese fleets were still at it, while species like the right whale and humpback nose-dived toward extinction. McIntyre began lobbying the International Whaling Commission, the multinational body that set hunting quotas, for a ten-year moratorium on all whaling. She wanted at least to hit the pause button, so that populations could be better studied and rebound. McIntyre had studying to do herself. She'd never even seen a whale.

Starting in 1971, McIntyre made a series of visits to laboratories and aquariums to learn about whale and dolphin intelligence. They were research trips—she called them "pilgrimages"—and among the most significant was her visit to the Communications Research Institute on St. Thomas. It was the headquarters of an Ivy League–educated neuroscientist named John Lilly who'd become enchanted with the enormous brains and intelligence of cetaceans while doing research for the U.S. military. (Among other things, Lilly was studying mind control, experimenting on dolphin brains as proxies for human ones.) Communicating with whales and dolphins became Lilly's life's work. He believed dolphins in particular could be coaxed to speak—to tune and shape their squeals into English words—and he and his colleagues saw themselves as working to usher in a harmonious "bispecies culture . . . of Dolphin and Man." Among many esteemed visitors to the lab was Carl Sagan, who felt that trying to talk to a dolphin was a prudent practice round for our first contact with extraterrestrials. Sometimes Lilly gave the dolphins LSD.

McIntyre was drawn to St. Thomas by a story that was, in a way, a famous precursor to the pseudo-marriage of George and Tex—another sensationalized case of animal-human cohabitation. In 1965, a young woman named Margaret Howe—a hotel worker on the island who'd shown up at the institute and started helping out—had the idea that creating a *need* for dolphins to communicate with humans might force a breakthrough, the way exchange students pick up languages abroad. And so she moved in with a bottlenose dolphin named Peter, and they lived together for two and a half months in a specially constructed "flooded house."

There was a kitchenette, an office with a chair and a desk, and a cot—all in twenty-two inches of water. Their daily schedule resembled that of a language-immersion preschool, with Howe patiently trying to teach Peter to say both their names, the numbers one to five, and so on. (She wore bright red lipstick so that Peter could better track the movements of her mouth.) They also watched television. "No matter how long it takes," Howe assured Lilly, "no matter how much work, this dolphin is going to learn to speak English."

Like George Archibald, though, Howe had essentially imprisoned herself with the animal and quickly experienced an agonizing loneliness. At night, she cried in her waterlogged bed. She complained of depression—also chafing. The dolphin, meanwhile, wouldn't leave her alone; he squealed over her voice with jealousy when she tried to talk on the phone, and he bruised her shins with his constant prodding. By week five, Peter was getting frequent erections, which kept him from concentrating on his lessons. So Howe learned to massage his penis, giving him an orgasm, so he'd calm down.

Ultimately, the scientific results were not illuminating. Peter the dolphin did not learn English. But living with an animal became an ascetic, spiritual exercise for Howe and her reflections on it were fascinating. (Her journal was included in Lilly's book, *The Mind of the Dolphin*.) Howe had ideas about how to tweak the experiment in the future, to make it more productive. "We owe it to the dolphin and to our curiosity to try it," she wrote. But Lilly, ostracized by the scientific establishment, soon lost his funding. In 1967, the Communications Research Institute closed. When Joan McIntyre arrived four years later, she found only squalor. Deformed sea turtles swam in the dolphin pool, and the hillside was strewn with trash. Howe was still living on the island, and McIntyre tracked her down. But Howe, disappointed by the way her relationship with Peter had been portrayed in the press, wouldn't disclose much. As McIntyre later put it, "Everyone saw her as a quirky animal-fucker."

In retrospect, the Communications Research Institute can be seen as one root of a new kind of empathy for wild animals in America— a feeling that Joan McIntyre would help carry off the island and back to the continental United States. Today that ethos subtly colors even mainstream conversations about what's at stake in wildlife conservation, and it was maybe even more mainstream back then. In 1972, for example, McIntyre helped bring whales into global focus as icons of the entire environmental movement—the polar bears of their day— by organizing a raucous march at a United Nations conference on the environment in Stockholm. There was a school bus gussied up to look like a whale and hippies handing flowers to the policemen they passed. But marching alongside McIntyre was former U.S. Secretary of the Interior Walter Hickel—a middle-aged Republican real-estate developer in a prim white tennis sweater, waving an American flag.

Two years later, in 1974, McIntyre published the holy book of this new whale-friendly worldview—an anthology pulling together many of the people and ideas she'd encountered on her research pilgrimages. It was called *Mind in the Waters: A Book to Celebrate the Consciousness of Whales and Dolphins.* Its pages exploded with native legends, scientific treatises on the anatomy of whale brains, beat poetry, briefs on whaling policy, and many drawings of naked people. Though it is obscure now, the book sold well and, as one historian puts it, was embraced as the *Silent Spring* of the burgeoning Save the Whales movement.

Publicizing the book, McIntyre emerged as a vibrant and unpredictable ambassador for the animals. At the time, one newspaper profile noted, most ecological crusaders were "edifying dullards." McIntyre, however, carried on about whales as "sensual, sexual animals," and about fostering more "whale connectedness"; for one photo shoot, she spontaneously French-kissed an aquarium's killer whale. ("I want to make love to a sperm whale!" she announced to a *Washington Post* columnist.) For her, the whale was "the guardian of the sea's unutterable mysteries" and the "deep calm mind of the ocean, connected to body, living *in* the world, not looking out at it." In short, the animals weren't innocents that we were obligated to protect. They were equals, maybe even superiors. McIntyre was decoupling the meaning of these animals from their ecosystem, warning that the most profound consequences of their extinction could be spiritual, not ecological.

Maybe this sounds stale and flaky—the stuff that gets garbled through a cloud of smoke in dorm rooms late at night. But that's just the point: in 1974, this was apparently a fresh and mesmerizing way to talk about wild animals. It was literally news, splashed onto the pages of the country's most straitlaced newspapers. McIntyre was making a novel argument for saving species, swirling the outrage and urgency of William Temple Hornaday together with the sentimentality of the nature fakers and dialing up the intensity of both. The result was sexy, heartfelt, explosive, and exuberantly unscientific—a vision of whales as shamanistic creatures who only "play, swim, look for food, love, eat, travel, take care of their young and each other," as she told one newspaper. "That to me is, like, a very instructive lifestyle. Their way could be an incredible teacher to us."

Since the dawn of modern wildlife conservation a century earlier, the threat to nearly every imperiled species on the continent seemed straightforward enough: people were shooting it. McIntyre was taking the stage at the beginning of a new era, when what menaced wildlife was suddenly oblique. In 1962, Rachel Carson had published *Silent Spring,* tracing how invisible pollutants like DDT were ricocheting around the environment, wearing away bird species such as the bald eagle, and infiltrating our own bodies. America seemed determined to solve these problems with nothing but more technology—by studying and tracking so many wild animals and atmospheric variables and pollutants that, eventually, we could trace the entire, baffling web of Earth's ecology and repair it. In 1970, when a female elk in Wyoming was outfitted with the first-ever satellite tracking collar by NASA and Smithsonian scientists, the contractor that had built the device, a company called Radiation Incorporated, promised in a press release that now only the "ability to measure a moving system" could "prevent extinction of animal species" and "help bring these menaces to human life under control." In an artist's rendering, the elk—nicknamed Monique the Space Elk—stood with vulnerable doe eyes, in a picturesque valley, while the electronic appliance around her neck ejaculated a thick thunderbolt to a satellite in the sky.

With whales, McIntyre had locked onto one of the last clear-cut conservation cases, where simply convincing humanity to stop hunting an animal could save it from extinction. But she also saw whales as guides to lure us away from the cerebral attitude toward nature on the rise. John Lilly, a neuroscientist-turned-conservationist, believed that if humans recognized an intellect like our own in whales, we would be moved to protect them. But intellect bored McIntyre; it was, for her, the root of the entire catastrophe, the very thing that had reduced whales to objects to kill and carve. Intellect built the exploding harpoon and industrialized the business of whaling, transforming little wooden boats into mechanized and efficient floating slaughterhouses. Intellect, she wrote in *Mind in the Waters*, had led us to look down on animals and "blinded us to their incredible essence." It had left us "incomparably lonely. It is our loneliness as much as our greed which can destroy us."

McINTYRE'S IMAGE OF PACIFIST, messianic whales may have been mostly a projection. But the fantasy was appealing, particularly against the backdrop of the Vietnam War. By the time *Mind in the Waters* was published, she'd formed a nonprofit called Project Jonah, and she and her two-woman staff found themselves backed by celebrities like Candice Bergen and Judy Collins, with a stream of unsolicited donations and lucrative invitations, like the one that brought her to the IBM sales retreat. Project Jonah organized rallies for children and flew three kids to Tokyo to deliver the twenty-six thousand letters and drawings they'd collected from children around the world, asking the Japanese prime minister to retire his whaling fleet. Momentum was building. But then, in the summer of 1975, Joan McIntyre watched the antiwhaling movement that she'd been nurturing stand up and walk off in its own direction.

On June 27, a fledgling band of activists from Canada rented a small fishing boat and intercepted a Soviet whaling vessel off the coast of Northern California, in a part of the Pacific known as the Mendocino Ridge. The group called itself Greenpeace. Until then, they'd been focused on the issues of nuclear weapons and power, but they'd learned that sperm whale oil, because of its low freezing point, was still used as a lubricant in the manufacture of nuclear weapons. It seemed like madness—one species driving another extinct in order to build a tool to extinguish itself.

Two of the men confronted the Soviet ship in a small Zodiac, situating themselves in front of a sperm whale like a human shield. The Russians fired anyway. A 250-pound grenade harpoon streaked over the men's heads and into the back of the whale. Those in the Zodiac stared into the animal's giant eye as it died. A second Greenpeace boat was filming the attack, and a few days later, Walter Cronkite somberly broadcast the footage to millions of Americans. The historian D. Graham Burnett writes that the event "galvanized broad and deep public revulsion at modern whaling, and modern whalers." It was a turning point: a century earlier, small bands of men huddled in little boats and risked their lives to hunt sperm whales; now they were ready to trade their lives to save one.

Burnett calls the confrontation "epoch defining." It introduced Greenpeace to America and unveiled a more theatrical, militaristic style of environmental campaigning—one that has compelled people to sit in front of trees when the loggers arrive, or lie down in front of developers' bulldozers, ever since. It believes that nature is worth dying for. And though it is absolutely radical in practice, in theory it seems to have a fairly wide appeal. (One of the young men in the Zodiacs that day was Paul Watson, whose splinter group, Sea Shepherd Society, known for ramming its boats headlong into Japanese whaling vessels, has been the star of Animal Planet's best-rated program since 2008, the reality show *Whale Wars*.)

This is to say, the Mendocino Ridge incident signaled the beginning of something powerful in America. But for Joan McIntyre, it was only an end. The episode sickened her. She worried that the tack Greenpeace had taken with the Russians could escalate out of control into an international incident, maybe even nuclear war. Having recently briefed one of Greenpeace's founders on the whaling issue, hoping to get the group involved, she now felt curiously betrayed. For her, saving whales had been about clearing more space for compassion in the world. Whales "were role models to increase our ability to get along with one another, and be affectionate, and re-establish our connections," she later remembered. "Now there was this heroic, macho event, where people are chasing Russian ships around and potentially catalyzing violence. That was the betrayal: taking something that I believed was gentle, and turning it into something warlike. Our natural ability to go wrong was manifesting itself."

It solidified something in McIntyre's mind. She'd been noticing that her efforts to increase empathy only backfired or fed old antagonisms. Among the letters to the Japanese prime minister she collected from kids, for example, she'd found more than a few attacking "the Japs" for being such barbarians. It seemed to her that good intentions were easily disfigured after too much exposure to the world, like metal corroding in the elements. There was something ugly and self-defeating in humans, or at least hapless, that always rose up to confound any progress. It was similar to how, ten years later, the crowd for Humphrey the Humpback would ravage a butterfly's habitat, and how a fence meant to protect the Antioch Dunes nearly tore them apart. (More recently, after I stopped visiting the dunes, I read that Fish and Wildlife had determined that herbicides it was spraying to control weeds there had probably been killing 25 to 30 percent of the butterfly larvae every year.) "All our efforts to control our future," McIntyre wrote, "are like ants walking around a Möbius strip."

She wasn't angry with Greenpeace. But she felt that whale activism had now shifted onto a new plane—one where she wasn't needed or useful. In 1977, she dropped the cause entirely. She moved to the small Hawaiian island of Lanai, where she lived in a tent on a cliff. She wanted to watch the whales—the actual whales—and did so with a childlike diligence that brought to mind Rudi Mattoni, withdrawing into his land in Uruguay to catalog the bugs. Before, McIntyre wrote, whales were "symbols flopping around inside my mind." Now, she "perceived the gigantic distance between the reality of the whales that I was watching from the cliff on Lanai and the effort to save them." Eventually, she moved to Fiji, where she married a native man, changed her name to Joana Varawa, and, after quietly publishing a couple of rambling memoirs, basically disappeared.

I'D APPROACHED McIntyre's story as one in a series, pulling apart the different ways in which America has thought about and cherished its wildlife over the last 250 years, from Thomas Jefferson's time, through William Temple Hornaday's, and up until my own. These aren't attitudes that switch on and off; instead, they've been draped over animals, and on top of each other, like translucent silk scarves, so that by now their patterns meld into something new and are no longer easily decipherable. So, having read so much about Joan McIntyre, I found it strange to watch the actual woman step off a ferryboat in downtown San Francisco to meet me one morning recently, nimble but worn-looking, with all that fire-red hair spilling from a faded baseball cap. It was as though she'd come hurtling out of history, to give me a hug and tell the barista her name—Joana—so we'd know when her americano was ready.

I'd tracked down Joana Varawa on Lanai, where she now lives again, working as an artist and writing an online community newsletter. She was back in the Bay Area, visiting her son and grandchildren for the first time in almost a decade. On the phone, she'd laughed when I asked if she still wanted to make love to a sperm whale, and laughed at what she called her youthful and naive belief that "if we could save the whales, we could save the earth"—even though that's exactly what Robert Buchanan had told me in Churchill about polar bears, and he believed it in his bones.

She told me right away that she didn't disown anything she'd done with Project Jonah, but she was disillusioned. The pointlessness and folly that she'd glimpsed at the Mendocino Ridge—"our species' absurdity," she called it—is only more apparent to her now. "The ultimate path in front of us appears to be the total extinction of everything," she said. But she was laughing when she said that too, and when I pointed that out, she shot back, "Well, because what else are you going to do?"

I was relieved not to find yet another crabby and wounded ex-environmentalist, and I asked her how she was managing to live in a world that she found so discouraging. The answer wasn't reassuring. She told me about the Taoists in ancient China. "They looked around and saw they were facing the same situation, a world that was disintegrating around them. And they realized the best thing to do is do as little as possible. Don't feed any new energy into a system that's falling apart, because you don't know what that energy will wind up being spit back as." Rather than try to change society, it's better to retreat. "You try to stay virtuous in your immediate life, you try to be correct—because you only feed the monster if you interfere too much." That was why she'd disappeared after the Mendocino Ridge, she said. She was done interfering.

Joana told me that she'd gone to one more International Whaling Commission meeting first, however. It was in Canberra, Australia, in 1977. The scene saddened her. Outside, she said, herds of new save-the-whales organizations were jockeying for money, influence, and attention; inside, nonwhaling nations traded their whaling quotas to Russia and Japan in exchange for increasing their fishing quotas. It was like a stock exchange to divvy up Earth's living things. The real turning point came when she met a Japanese delegate who told her that he now spent more than two hundred days of every year flying to and from fisheries meetings just like this one. She looked deep into his exhausted eyes and saw a man trapped in needless and hopeless complexity. She pitied him, she told me, and as she described it, I couldn't help think of the much smaller squabbles flaring in the Whooping Crane Eastern Partnership, among all those people who wanted only to save a beautiful bird. "My original intention was to soften things," she told me, "and this was just all getting so hard."

So she left. Early the next morning, she bought up all the daffodils she could find in downtown Canberra, and before anyone arrived for that day's meetings, she placed a few on every conference table. Then she went to the airport and flew home. "And that," Joana said, laughing, "was my last act as a whale saver."

AROUND THANKSGIVING, Operation Migration caught a streak of good weather and rallied. The cranes were locking into the trikes and had built up enough stamina that, during some flights, the pilots found they could skip a stopover and keep pushing the flock to the next one. During one eight-day stretch, the team flew on four different mornings, shredding through seven migration stops and 350 miles, and catapulting the cranes out of Kentucky, through Tennessee, and across the Alabama state line. There was talk of finishing the migration before Christmas, which hadn't happened since 2004.

Still, there was one ongoing headache—crane number 2, the same bird that had freaked out and retreated into the woods that first morning in Necedah. Shortly after, the crew discovered that the bird had sustained an injury, a nasty abrasion along the leading edge of its wing—it was ripped up, with feathers and fibers dangling from the flesh. After the wound was treated, number 2 seemed to recover quickly; soon it raced out of the pen in the morning with the rest of the gang and thrust out its wings as though it were going to spurt into the sky. But it seldom left the ground. And when it did get airborne, it bailed out after a quarter-mile at most and had to be tracked down on foot by OM's costumed ground crew. Often the bird would land deep in the woods, or far off the road in a gauntlet of farm fields and six-foot-deep irrigation ditches, and the ground crew would have to hike in after it, then hoist it in a wooden crate on their backs, all the way back to their van. (Experience shows that a whooping crane can be driven a few legs of the migration and still learn the route, though it's unclear how many exactly.) One morning, number 2 did lift off from the pen, but then rocketed straight into the top of a tree, ricocheted down through the branches, Pachinko-like, and landed with its legs wrenched open in a V.

For weeks, this one problem child siphoned up a tremendous amount of Operation Migration's time and patience. By the time they'd reached Tennessee, number 2 had become the subject of many internal e-mails and conference calls between the WCEP partners. Almost everyone in OM was convinced that the bird was incapable of flying—that the torn wing had been more serious than anyone realized. But the scientists at Patuxent argued that number 2's problem might only be psychological, that the bird had been traumatized and needed to work through a fear of the trike. One government researcher told me that Operation Migration seemed to be giving up on the bird too easily: "My attitude was: Try harder. This is what you're there for—to deal with the difficult ones, not just the easy ones." In the end, the partners consented to have the bird driven to a vet in Nashville.* After an X-ray showed that number 2 had, in fact, ripped a membrane in its wing and would never fly normally again, the bird was taken out of the migration and sent back to Patuxent. But by then, the whole debate about the bird between Operation Migration and the other partners seemed to have dragged out and gone sideways, in that special, galling way that disagreements between unhappy spouses do.

The intensity of those conversations, at least on Operation Migration's side, was probably amplified by a more serious disagreement developing that fall. During a parallel series of conference calls, a team of WCEP members, including a representative from OM, was hashing out a new long-range plan for the partnership. And by now, WCEP's commitment to putting its work back on firm scientific footing was presenting an existential threat to Operation Migration. The pilots were being argued out of their jobs.

WCEP's initial goal had been to release a new troop of whooping cranes at Necedah every fall until it had built up a self-sustaining population of between one hundred and 125 birds—whooping cranes that were nesting and breeding at Necedah at a healthy rate and autonomously teaching their young the way to Florida. Only then could the humans stand down.

With a hundred birds now soaring down the flyway, WCEP was extremely close to reaching that goal. Many of these cranes did nest

---

* Taking a whooping crane to the vet on migration is not like taking a house cat to the vet. Someone first had to go ahead, Secret Service–like, to secure the clinic. Barking dogs were cleared out. Phones were silenced. The doctor was outfitted with a hood and white costume. Actual local headline: "Nashville Vet Dresses Like a Marshmallow to Save Whooping Crane."

and mate in Wisconsin. But they routinely abandoned their eggs. After ten years, only one chick had survived long enough at Necedah to follow its parents to Florida. The discouraging truth was that, although large numbers of whooping cranes had been miraculously conjured into a place where, previously, there'd been absolutely none, those cranes weren't yet taking care of themselves. Whooping cranes would continue to live in Wisconsin only if humans kept breeding them and putting them there. That is, the reintroduction was working fantastically. But, looked at in a narrower and less forgiving way, it was still a near-total failure.

And so, earlier that year, a high-ranking panel of government scientists and conservationists known as the International Whooping Crane Recovery Team decided not to allow any more whooping crane chicks to be trained for migration at Necedah until the breeding problem was studied exhaustively and solved. (One theory was that an infestation of black flies was biting and hounding the cranes until they were forced off their nests. But there was also vague concern in the partnership that the problem could be less straightforward: that, maybe, being reared by a big white costume, instead of a bird, had written some inscrutable glitch into the programming of these cranes, making them inadequate parents.) Pledging more animals to that stalled population seemed unwise at this point—like potentially throwing them away. Operation Migration could shift its base of operations to another site next spring, and still lead a new group of cranes on migration. But that meant zeroing in on a piece of suitable habitat, evaluating that land with a battery of scientific tests, then getting the requisite pile of permits completed before next spring—a ruthlessly tight schedule. And many in WCEP were adamant that they'd rather miss the deadline, and leave Operation Migration behind, than rush into anything haphazardly.

There were now also alternative ideas for teaching cranes to migrate surfacing in the partnership. (For the last several years, for example, the International Crane Foundation has been releasing a small number of juvenile, captive-reared cranes near the adult whoopers already at Necedah and coaxing them to follow those older birds south instead of an ultralight. In short, they're starting to succeed in doing what they'd once done with sandhill cranes, but had seemed impossible to pull off with whooping cranes at the outset of the reintroduction.) For some in WCEP, the breeding deadlock at Necedah cleared a perfect space in which to pause and dedicate more resources to investigating these new techniques. And in certain corners of the partnership, there was only low-grade sympathy, at best, for OM's predicament. John French, Patuxent's research director, told me: "Decisions should be based now on data. They need to be made for the good of the project and the good of the birds. If that means that OM isn't part of the project anymore, then that's the way it goes. So be it."

By late November, with the crew's pace slowing through central Alabama, finding and vetting a new piece of land in time to launch an ultralight migration the following year started to seem impossible. One night, scooping leftover barbecue onto a paper plate in a host family's kitchen, Joe Duff pulled me aside and admitted that he was all but convinced this would be Operation Migration's last migration. If they were forced to stand down next year, they could never rebound: their funding would dry up, and the crew would take other jobs. He felt bad that he couldn't go public with that news yet, he told me. He would have liked to give the many Craniacs along the route a proper goodbye.

As improbable as it sounds, there was actually a second, smaller band of costumed whooping crane conservationists barreling south behind us in campers and vans that fall—a kind of shadow operation, a couple of states back but gaining ground. This was WCEP's tracking team.

The tracking team, primarily staffed with young biologists from the International Crane Foundation, has been charged with monitoring all the other birds in the population—the one hundred whooping cranes now migrating south independently, and especially those making the trip on their own for the first time. Every crane has a transmitter around its leg that spits out a blip on a personalized radio frequency. The trackers drive ahead of the birds, plotting their positions and progress as they go. Early in the morning, they stake out the cranes' last known locations to try to get eyes on the birds, and they collect reports from a network of bird-watching cooperators in various states. They do this all fall, until the birds get to Florida. Then, in the spring, they do the whole thing again, in reverse, chasing the birds on their migration back to Wisconsin.

Sometimes there is a hands-on element to the trackers' work. If a bird winds up somewhere it shouldn't be—a playground, a hunting ground, hundreds of miles off course on the opposite side of Lake Michigan—the trackers are ready to flush the bird, or even suit up in a white crane costume, grab the crane, and drive it somewhere better. You can imagine the trackers as white-robed guardian angels, watching over the cranes and rescuing ones that get into trouble. Or you can see them as a roving goon squad, abducting and reeducating any birds who don't follow the rules.

That ambiguity exists inside many other endangered species recoveries, too. Around America, wildlife managers are laboring to negotiate the same sorts of impossible boundaries between animals and the human world, or even between animals and other animals—policing their wildness. Bison that wander out of Yellowstone Park have been pushed back across the line, for fear they'll transmit diseases to livestock. When red wolves were reintroduced in North Carolina, the animals were outfitted with special collars that could be triggered remotely to stick the wolves in the neck with a sedative. It was a kind of kill switch, in case the humans who were monitoring the wolves saw an animal's signal wander outside a certain designated territory but couldn't get on the ground quickly enough to intercept it. In other parts of the country, if reintroduced gray wolves can't be kept off ranchland and they kill a rancher's cow, state governments or conservation groups have simply reimbursed the rancher for his losses.

Wildlife officials have used paintball guns to push moose out of traffic on Interstate 90 in eastern Washington. In Alaska, they use Tasers. In Texas, as part of its reintroduction of rare desert bighorn sheep, the state has sent out marksmen to gun down preemptively the animals it sees as competitors to the bighorns, clearing feral donkeys and elk out of their way, as well as another, non-native species of sheep called the aoudad. Since 2001, a division of the U.S. Department of Agriculture, called Wildlife Services, has shot an average of six hundred coyotes a week from helicopters to clear western lands for domesticated sheep or cows or to protect mule deer. And near Cape Cod, where summer feeding grounds of the world's last 350 North Atlantic right whales coincide with a busy shipping lane, with boats charging in and out of a liquefied natural gas terminal, Cornell University has installed a network of high-tech buoys to listen for the animals. Ten-second audio clips of possible whale calls are uploaded to a lab at the university, where analysts, working round the clock during the busiest times of year, verify these calls and warn any ships in the vicinity to slow down.

Zoom out and what you see is one species—us—struggling to keep all others in their appropriate places, or at least in the places we've decided they ought to stay. In some areas, we want cows but not bison, or mule deer but not coyotes, or cars but not elk. Or sheep but not elk. Or bighorn sheep but not aoudad sheep. Or else we'd like wolves and cows in the *same* place. Or natural gas tankers swimming harmoniously with whales. We are everywhere in the wilderness with white gloves on, directing traffic.

In an environmental law review article on the subject, the scholar Holly Doremus criticizes this sort of restrictive wildlife management as "erect[ing] a virtual cage" around animals. Conservation should strive to build healthy populations of animals that are "as wild as possible in a tame world." But controlling and manipulating animals, or making apologies and payouts for them, effectively amounts to a kind of domestication, changing species "from wild animals to human creations designed to serve human needs." Just as a cow on a fenced-in pasture meets America's need for beef, a buffalo scrupulously contained inside a national park meets America's need for buffalo: perhaps just our aesthetic and moral needs to know that buffalo exist there, and to keep their destruction off our conscience.

This work goes more gracefully at some intersections of people and animals than others. I learned about one management scheme that failed so badly that, when some environmentalists talk about it now, it can sound like an allegory or a farce. In 1987, the government decided to relocate 140 Pacific sea otters, a species once believed to be extinct, from their last redoubt on the Northern California coast to the waters around an unpopulated island far to the south, near Santa Barbara. (As with the eastern migratory population of whooping cranes, the idea was to seed a second, discrete population of otters, as a catastrophic insurance policy.) Not everyone in Southern California welcomed the otters, however: commercial fishermen saw them as competitors for abalone and urchin, and the oil industry worried they'd complicate its plans to drill nearby. So those engineering the otter recovery cut a deal. They drew a line around a large swath of ocean and promised that, if any of the otters left the island and penetrated that area, the government would trap them and move them back. The area stretched three hundred miles down the coast, all the way to the Mexican border. People called it the Otter-Free Zone. Essentially, it was zoned for fishermen but not sea otters, just as a downtown neighborhood might be zoned for light-industrial use but not residential.

Most of the otters were affixed with radio tags, but not very effective ones. The signal couldn't be detected more than two miles away, and the batteries ran out after ninety days. (A few otters were outfitted with more sophisticated tracking devices, but those had to be implanted surgically.) And so, a telephone hotline was set up; the public would report any otters they saw on the wrong side of that invisible line. Over the next several years, a biologist named Greg Sanders drove around Southern California responding to calls. He would arrive on the scene, often days later, and stare into kelp beds with a telescope for hours. Sometimes the otter turned out to be a sea lion or only a floating log. Often Sanders found nothing. There was an otter-in-a-haystack feeling to the work.

After six years, twenty-four trespassing otters had been captured and removed from the Otter-Free Zone. Each was driven or flown in a chartered plane to a release point hundreds of miles north, at a reported cost of $6,000 to $12,000 per otter. At least four otters died after being moved, probably from stress. Two others immediately swam back to where Sanders had trapped them. In 1993, the government finally gave up enforcing the Otter-Free Zone. Sanders told reporters, "The underlying message is, otters don't stay where we put them."

ORIGINALLY, THE POSSE of whooping crane trackers was assembled by WCEP to do the same job as Greg Sanders: make sure that wild animals respected certain legal agreements that had been brokered by human beings—and which, of course, the birds could have no conception of. Before WCEP could let any whooping cranes loose—to fly and land wherever they pleased, potentially carrying with them some of the protections and restrictions associated with endangered species—the partnership had to negotiate with twenty different states in and around the proposed flyway, reassuring them that it could retrieve any birds that touched down where they weren't wanted. In ten years, relatively few birds have had to be relocated. But what's clear in the stories of the ones that have is that the trackers aren't just policing the law. They're policing an aesthetic—making sure that the whooping cranes we have created behave like the aloof and magisterially wild ones in our imaginations.

The wildness of an individual whooping crane is a fragile, slippery commodity. In the run-up to the reintroduction, scientists experimented with actively conditioning the cranes to avoid people by waving shiny balloons at them, running at the birds while screaming or rapidly opening and shutting an umbrella. ("Firing shell crackers during a charge is also helpful," one scientific report noted. "Less than helpful has been our few attempts to use chemical mace, an electrical cattle prod, pepper spray, and lemon oil spray to promote wildness.") In other words, there was an acknowledgment, right away, that the costume doesn't *make* the birds wild; it only keeps them from getting inadvertently tamed by prolonged exposure to the person hiding underneath it. Ultimately, every crane decides for itself what distance to keep from humans when, after learning the way to Florida and striking out on its own, it's finally allowed to encounter us unmasked. And some birds, despite WCEP's best intentions, have exhibited a distressing nonchalance.

Whooping cranes have turned up in suburban backyards and trailer parks. One liked to roost in a retention pond near a Walmart. One was seen strolling into a barn. Another male kept landing inside a zoo in Florida, drawn to the captive female whooping crane that lived there; once, the WCEP trackers had to fish him out of the zoo's bear enclosure. In the summer of 2009, another male crane became a fixture at an ethanol plant near Necedah, feeding on vast piles of waste corn, oblivious to the truck traffic grunting in and out. That bird was eventually yanked from the population entirely. Its wildness had been compromised, and it was a bad influence, luring other whoopers to the ethanol plant with it. It lives in a zoo in Tampa now and goes by the name "Kernel."

Cases like these have forced the question within WCEP of what "wild" even means—what should a population of wild whooping cranes look like in twenty-first-century America? It would be easy to say that a wild animal is one that lives outside human influence and beyond human contact—an animal that doesn't notice us or give a damn. But in that case true wildness would be almost impossible, with human influence now bleeding into virtually all the available space. Even WCEP's trackers don't have a written definition of wildness, or any empirical metrics to enforce it with—how far away from a Walmart a wild crane *should* stay. Ultimately, wildness is a matter of individual opinion, and not even the experts agree.

Inside WCEP, there are wildness fundamentalists and those who take a more libertarian tack. When I visited Patuxent one summer, John French, the research director, told me that lately he's poked at this confusion in the partnership—the squishiness of what wildness is—by asking his colleagues to do little thought experiments. Sup-

BIRDS

pose, for example, that the partnership managed to establish a large and healthy population of whooping cranes in Wisconsin, with twenty-five or even fifty breeding pairs, each fledging two chicks every year. But suppose those birds nested on golf courses and frequently had to be hazed off the back nine, like Canada geese. On a visceral level, French said, this would feel like a disaster. The romance of the whooping crane is bound up in its being not just a wild animal, but a *very* wild animal, he told me. That wildness drew America to the whooping crane's cause in the first place. And after such a long fight, "who wants to see whoopers wandering around a parking lot eating French fries? I certainly don't." But French also said he might have to force himself to accept such a scenario as success.

Our vision of wildness may be impossibly nostalgic, an almost religious fantasy of purity in what's remote, in what's beyond us—not unlike the gentle deities that Joan McIntyre saw in whales. It may be unfair to expect actual whooping cranes in the twenty-first century to behave the way we imagined whooping cranes did in the sixteenth century. In a world full of Costco regional distribution centers and Krispy Kreme drive-thrus, we are asking them to block it all out, to see the Walmart retention pond as a slum instead of a providential new form of habitat in a changing world, and to see the corn piling up outside an ethanol plant not as food but as "waste product," and to decide, as we have, that eating it is beneath their species' dignity. Maybe we want the cranes to be anachronisms, to live in an avian version of Colonial Williamsburg, by the code of their ancestors, and without whatever tools the modern world might provide for them.

French conceded that the entire *Truman Show*–esque existence that's been concocted for these cranes in pursuit of that wild ideal may not even be the best thing for the birds. Let's face it, he said: "The animals that survive and thrive are those that do well in and around a human-oriented landscape." The purest triumphs of conservation may be the species that rebound so phenomenally that they become nuisances to us on our own turf, wiggling off wildlife refuges and into niches that we didn't expressly clear for them. Think of the white-tailed deer, an animal William Temple Hornaday presumed to be doomed in the Northeast a century ago in *Our Vanishing Wildlife.*

The same story has played out to a less exaggerated degree around the country with cormorants, pelicans, and sandhill cranes. Or consider Canada geese, one subspecies of which—the giant Canada goose—was once even presumed extinct in the United States, until a single remnant flock was discovered in 1962. Panicked scientists held scientific symposia focused on the birds' preservation and, even as late as the 1980s, geese were being reintroduced around the country, and fed rations of corn through the winter by state governments. Now there are 3.5 million geese in America. They're so numerous that they get sucked into jet engines and take down commercial flights, and the federal government rounds up and gasses untold thousands every year.

Even pigeons were once cherished in American cities, before all the handouts and garbage we've given them to eat allowed their numbers to explode. In 1878, the *New York Times* described pigeons as "honest birds" whose "right to feed in the street" was being challenged by sparrows. In the early 1930s, a community of gentlemanly pigeon feeders in front of the White House hoped that President Franklin Delano Roosevelt might come out to join in their "highly developed art."

It's hard to square our nostalgia for certain rare species with our resentment of species, like these, that we've helped to thrive, intentionally or unintentionally. It's a thin and erratic line we draw between the wildness that awes us and the wildness that only annoys us. It's a reminder that we remake the animal landscape on timescales longer than our imaginations are calibrated to perceive or predict, and that we can't predict how we'll feel about those changes, either. At Antioch Dunes, I'd been unsettled by how shifting baselines syndrome cloaks the past, warping our understanding of what's been lost. But it hadn't occurred to me to project the problem of shifting baselines into the future, too, to wonder where what we're building and preserving is heading, and if it will be judged differently by those who inherit it.

A root of the problem appears to be that, even though empirical targets are typically laid out in the government's recovery plans—*x* number of whooping cranes, and so on—the people steering those recoveries have been working so hard, for so long, to prevent the chastening failure of the species' extinction that they can't quite envision what success should actually look like, in the real world. (As John French told me at Patuxent, "At this point, work on any endangered species is certainly a very severe rearguard effort. In our meetings in WCEP, we don't talk about underlying philosophies of humans and the natural world. We talk about what we're going to do next week, and who's got enough money to buy food.") But there's also a fickleness to our attitudes about wildlife over time that I hadn't acknowledged. And, like the problems of shifting baselines and climate change that I'd come across before, it creates one more cavernous and disorienting ambiguity at the heart of conservation. America is scrambling to save so much. But, as Holly Doremus writes, we've never asked "how much wild nature society needs, and how much society can accept."

I DECIDED NOT TO TAKE Isla on any of the migrations that fall. Partly, it was because I didn't want to subject her to the miserable logistics of crane chasing—to shlep her across the country to some deep Southern backwater motel and wake her before dawn four or five mornings in a row, in case the wind happened to blow in precisely the correct direction, at a cooperative clip.

I remembered, as I'd stood by at Necedah, waiting to see whether

Operation Migration would finally lift off, watching the defeat expand in the eyes of the Japanese television crew's cameraman—how tired he looked, and how, over the course of that week, his face gradually surrendered to a beard. I had mixed feelings about inflicting that hassle on my little girl. And I wasn't sure it was worth it. By now, the momentary sight of a few white birds flapping behind a plane at a flyover event almost felt secondary. Whatever story I was following at this point wasn't about the breathtaking inscrutability of animals, but of people.

I do remember, though, how, early one morning when Isla was three and a half, she and I caught sight of three raccoons outside our kitchen window. A fat one right in front of us, the size of a small schnauzer, scampered back and forth along the wall of our little cement backyard. Two others—one big, one small—clambered up the posts of our neighbor's deck. Then down again. Then up.

"Baby raccoon!" Isla said in an explosive stage whisper. We watched, rapt, as they hurried to do whatever raccoons try to get done before sunrise. They were Isla's first raccoons.

I like raccoons. I can't understand why they're so underappreciated—not detested like rats or opossums maybe, but not known for delivering thrills, either. They're surely one of the most cuddly looking and admirable synanthropes—what biologists call species that succeed in the human environment. (In the last seventy years, the number of raccoons living in North American cities has grown twentyfold.) I'd read that President Coolidge's family kept a pet raccoon named Rebecca at the White House. A 1963 essay in *Harper's Magazine* titled "Our Most American Animal" even proposed that the raccoon be elevated alongside the bald eagle as a second national symbol. The raccoon's character, the essay argued, is our character. The animal embodies everything that separates the New World from the Old: toughness, adaptability, self-reliance, a devotion to being busy: "Nothing about him is rare, delicate or specialized. He is as common as dirt and as hardy as weeds."

When the raccoons moved on, Isla asked me to pour her a bowl of Cheerios. Our day started up as usual. She rarely mentions watching polar bears in Churchill. But from time to time, she asks me if I remember that "very special time we got to see raccoons."

14.

SPOILER

L ate one morning in early December, Operation Migration touched down on a hay farm in Chilton County, Alabama, smack in the center of the state, at the migration's 813th mile. As Joe came in for a landing, a retired veterinarian living on an adjacent property, hearing what sounded like some kind of wailing industrial leaf blower, rushed to his window and was stunned by the sight of a hooded figure in the open-air cockpit of an ultralight, descending with a line of white birds snapped behind his wing, like an elaborate kite in a stiff wind. The veterinarian's name was Bubba, and he wasted no time ordering up dinner for everyone. By nightfall, trays of chicken, ribs, sweet potatoes, and turnip greens were spread like a church buffet across a table in Bubba's front yard. Slabs of homemade peanut brittle were passed around the fire pit.

By then, Operation Migration was zipping toward Florida at a phenomenal clip, weeks ahead of where they were the previous year. But a heaviness had set in. Layers of crankiness had accumulated as we rolled south, the way a ball of ice grows, careening downhill. Or at least that crankiness was being shared with me more candidly. I noticed that now, when the pilots met at dawn to assess the weather for flying, there was a subtext to the conversation, in which each of the three men passive-aggressively angled to be the one who made the correct call. Even Gerald, a breezy older fellow on the ground crew who cooked up big egg breakfasts on down days and curates a section of OM's Web site called "Gerald Murphy's Recipes," started dumping on me all of a sudden when we found ourselves waiting for a few minutes alone in a parked car. He felt he was being "taken for granted," he said, and was contemplating volunteering instead for a red-cockaded woodpecker conservation project near his home in Florida. "I miss sleeping in my bed with my wife," Gerald told me glumly.

But it was Brooke Pennypacker, one of the pilots, who seemed to be having the toughest migration. Brooke was sixty-one that autumn—the same age as Joe—a wry New Jersey native with downy, chalk-white hair and large stores of conversational energy that rarely found an outlet, since he worked with birds he was not allowed to speak near. Some coworkers had taken to calling Brooke "the Vortex," because he was known to suck up their time unexpectedly with long and extemporaneous stories. But I'd been drawn to the guy since my first day at Necedah, finding those stories to be invaluable peeks into the wackiness and dreariness of daily life on migration—of what it actually means to have devoted oneself to a bird.

Brooke's lifestyle was more nomadic than even his fellow crew members': He essentially migrates with whooping cranes year-round, helping to raise each new class of birds from the time they hatch at Patuxent in the spring, relocating with them to Necedah for flight training in the summer, maneuvering the cranes south with Operation Migration all fall, and staying behind in Florida to monitor them through their first winter. And by then, it's nearly time for him to start the entire cycle over again. "Brooke's spent every waking hour on these damn birds now for seven years," OM's ground crew chief, Walt Sturgeon, told me. "That dedication is unbelievable. But he's getting very cynical, and that's too bad. Then again, I don't think there's anything I can do to stop it or change it." By the time OM reached Chilton County, I'd noticed that the stories tumbling out of Brooke were mostly circling back to the same place: some bemused or defeated complaint about WCEP, or about the other two trike pilots. Usually, his ranting was good-humored and self-deprecating. But lately, when he started, I noticed other crew members slowly inch away.

Grounded at the hay farm, Brooke and I managed to kill most of a morning talking in the run-down barn. He explained how he'd wound up in this morass of crane politics—it was by accident, and for reasons similar to mine. He did it for his child. Brooke has a teenaged son, Devin, whom he'd raised mostly on his own for the first five years of the boy's life. Devin was about three and a half, in the mid-nineties, when Brooke met Joe Duff at a barbecue for ultralight pilots. Joe invited Brooke and his boy to visit him at the Airlie Center, a bird sanctuary and conference center, not far from where Brooke was living, in Virginia. (Joe and William Lishman had led geese to Airlie on Operation Migration's early, experimental migrations.) Brooke and Devin showed up first, and Brooke sat his son on the slope of an earthen dam so he could take a closer look at a few geese he heard honking in the nearby woods. When he turned around though, he saw a much larger squadron of geese treading over the top of the dam, moving with clunky determination, straight at Devin. Brooke froze; not knowing anything about geese, he worried they'd peck his son open like a bun. Instead, the birds only settled all around the little boy, put their heads down, and fell asleep. Brooke told me his son was vibrating and grinning, suppressing a squeal. "And I said, 'Shit, we've got to do more of this.'"

It wasn't that Brooke was a conservationist, or even interested in birds. But he felt compelled to engineer a certain kind of childhood for his son, just as I did for Isla—one that involved being outdoors, knowing about animals, heels dug into nature. "That's our responsibility," Brooke told me. "You've got to expose them to as much of the good as you can."

So, when the Airlie Center started a project to reintroduce trumpeter swans using ultralights and needed a pilot, Brooke took the job and moved onto the property. (Brooke would eventually join Operation Migration in 2002, after one of its original pilots had a stroke.) As his son got older, Brooke put him to work at Airlie on weekends, dragging him out of bed in the morning and into a kayak to haze any swans that tried to land in the sanctuary's pond during flight training. In a book about the swan project, I found a photo of Devin, then age eight, scrubbing out a swan pool with a long-handled brush. After a while, the kid started to resent it. "He had a lot of years of that," Brooke told me, his voice nearly dropping out from under him. I assumed there'd been a falling out—a fracture of some kind. But, for once, a tangent had arisen in a conversation that Brooke Pennypacker declined to go off on.

Eventually, we wound around to the story of another family. In the fall of 2006, three whooping cranes—two graduates of the ultralight migration and their new offspring—flew from Necedah to the Chassahowitzka National Wildlife Refuge in western Florida for the winter. But a drought had dried up much of the adults' traditional territory there, so they shifted to nearby Tooke Lake, ringed by neighborhoods of ranch homes, a golf course, and a trailer park. As the winter wore on, the three whooping cranes started eating from bird feeders in one particular backyard. Soon they were spending nearly every day at the house, pecking away, unfazed by the bird watchers and neighbors who came to gawp at them up close. When the WCEP trackers knocked on the door of the house, the old woman who lived there refused to take her bird feeders down. One tracker told me, "She was a little bit crazy, a little confusing to deal with. I don't know if she lives alone or what. But no matter what you told her, she was going to feed those cranes."

It was a complete breakdown of whooping crane wildness. And it stung especially because these three particular birds were supposed to be WCEP's big success story. They were known as the First Family. The juvenile was the only chick in the history of the reintroduction that had not been abandoned at its nest and had survived long enough after hatching at Necedah to make a migration with its parents the following fall. It was, in short, a celebrity—the only whooping crane in the eastern United States that had lived its entire life outside a lab, reared by other whooping cranes, not a human in a white gown. It was unimpeachably wild. But now its parents had led it here and taught it to scratch seed from an old woman's hanging plastic tube, like a common sparrow. The humans hadn't failed. The cranes were undermining our ambitions on their own—they were like the humpback that, having been laboriously rescued and sent back to sea, swam back and beached itself.

All of a sudden, the philosophical debate about wildness within WCEP had a focal point. Some saw the birds' behavior as an unbearable black eye on the project. But George Archibald, for example, visited the woman's house several times and saw no problem with the arrangement. Once, George told me, he watched the three whooping cranes pass only a few feet away from some kids fishing at the edge of Tooke Lake, without any sign of alarm. The kids ignored the cranes, and the cranes ignored the kids, George remembered, and that struck him as the best humanity could realistically hope for. "The purists want these birds to be wilder," he said. "They don't want them hanging around bird feeders or houses in Florida. Well, I'd prefer them to be that way, too. But we live in a fallen world. People are all over the place." George's International Crane Foundation, one of the WCEP partners, even capitalized on the situation by taking some of its donors to Tooke Lake to show off the First Family at close range.

Brooke visited the house himself late that January, he told me, arriving in street clothes to watch, from an adjacent empty lot, the three cranes strut across the old woman's yard in plain sight and snack on birdseed. It was hard for Brooke not to take insubordination like that personally, having spent so many summers roasting inside his costume during flight training, supposedly so the cranes wouldn't feel at ease around humans.

Suddenly, the old woman appeared in her housecoat. She was defensive and argumentative, Brooke remembered, and had already had enough of whooping crane conservationists showing up at her door and telling her what to do with her bird feeders. (I couldn't help picturing one of those crazy pigeon ladies, dispensing seed from a collapsible grocery cart on city sidewalks.) Just as the woman insisted to Brooke that she was doing nothing wrong, he told me, a wild sandhill crane chick came sauntering around the side of her house, stopped squarely at her feet, and looked up at her like a lapdog begging for a treat. Brooke's eyes bugged out. "What's that?" he asked her. "My little sandhill," she told him defiantly.

Brooke left, convinced that something needed to be done. But a few nights later, a massive storm surge battered into the whooping crane pen at the Chassahowitzka National Wildlife Refuge nearby. The newest class of cranes, which Operation Migration had just led to the refuge from Wisconsin, were still being penned every night while they acclimated to their new habitat, and seventeen of the eighteen birds drowned that night and died. There was speculation that the cranes were stunned by a lightning strike in the water just before the biggest waves hit. One of the carcasses was found with its talons stuck in the fence, as though the bird had been trying to climb out of the flood. Within days, the eighteenth bird was found dead elsewhere on the refuge, starved to death or killed by a predator—it wasn't clear. Just like that, an entire year of work was gone. It was the starkest, cruelest setback in the history of the project. Everyone set aside the more nuanced, philosophical setback playing out in the old lady's backyard.

LATE IN THE FALL, during the whooping crane migration—it was almost winter by then—I visited the woman on Tooke Lake. Her name is Clarice Gibbs, and she lives in a low-slung prefab house on the lake's north side. She has waves of short white hair and dark eyebrows, and speaks haltingly, searching for the right words—in short, not at all the defensive and muttering old bird lady I'd expected. On the phone, she'd sounded touched and relieved that someone finally wanted to hear her side of the story.

Gibbs told me that the whooping cranes hadn't been back to her house in years, but her backyard was still a playhouse for all kinds of birds: cardinals, blue jays, chickadees, titmice, woodpeckers, zebra finches, the occasional wood stork. She has been feeding birds ever since she and her husband moved there, she said, some twenty years ago, and I counted more than a dozen bird feeders in her backyard, of all different shapes and colors, hung from trees or hooked to stakes in the grass. A wooden block coated with peanut butter was suspended in a metal cage—a trick she'd read about in a specialty magazine called *Birds & Blooms*. "Birds are very entertaining," she kept saying. "We just love the wildlife."

Gibbs offered me a seat at her dining room table and showed me snapshots she'd taken of the First Family. I subsequently learned that there had been whooping cranes at her house, at least occasionally, during two other winters, as well. (One of the others was crane number 710, the whooping crane that, having acquired such a fearlessness of humans in her yard, returned to Wisconsin the next spring, became a fixture at the ethanol plant in Necedah, and eventually had to be exiled to a zoo.) But Gibbs was having a terrible time that afternoon reconstructing the chronology for me. She strained to sort out how many different birds and different winters that she was remembering. She squinted to read the dates printed in the bottom corner of her photographs for help. Frequently, she removed her glasses so she could just rub her temples and think. "You know, Jon," she said finally, "we—I lost my husband this year, so my dates aren't great. Sometimes I don't get everything together the way I should."

It seemed to her that WCEP's trackers started knocking on her door as soon as the cranes arrived. "And I told them," she stammered, "I said, 'We are not feeding those birds.'" By this, she meant that she was not *deliberately* feeding the whooping cranes. They were eating food that she put out for other birds. She and her husband knew all about the crane reintroduction, she told me. Everyone did; the men with the airplanes make the local news every winter when they arrive at the refuge with a new group of birds. "I read enough about them in the paper to know that they're trying to train them to migrate back and forth, and they don't want human contact," Gibbs said. So she and her husband tried to cooperate, only watching the cranes from their porch and never going into the yard or approaching them. For the most part, she added, the trackers "were very nice, very informative. And I could tell that their feelings for the whooping crane are very deep. They want to protect them, and I do, too." She wasn't trying to be nasty toward them or disrespectful. "And when the whooping cranes left, it didn't bother me at all, because I knew they were supposed to go. They weren't supposed to be here in the first place." But she couldn't afford to lose all the other birds along with them, even temporarily—and that's what would have happened if she took down her feeders. She knew this from experience: she and her husband had taken several long trips, a month at a time, and it always took several months of feeding after they returned to repopulate their backyard with birds. Gibbs then said something that might sound obvious, but she said it very slowly, because it was important that I understand: "If you stop feeding the birds, they stop coming. And you don't get to see them anymore."

She took a breath, then suddenly stiffened up. "My husband had Alzheimer's," she told me, "so our porch was a big part of our life." By the time the First Family arrived, she and her husband had done all their traveling—I got the impression that those long trips had been a kind of farewell tour to see his family—and were settled in at home, prepared for his condition to get worse. "A big part of our day was watching them," she said—drinking iced tea on the porch together, noticing the funny things birds did, and the cliques they formed. It was like people watching, but with birds.

Gibbs felt bad if she'd spoiled any of the conservationists' hard work, but she wasn't apologetic. After all, she saw herself as being 100 percent on their side. She was grateful to these men and women who'd worked so determinedly to bring the whooping crane back, because, aside from all the usual reasons, they had made it possible for one of those birds to alight in her backyard and bless what were effectively some of her last real days with her husband. He was aware enough to get a big old grin on his face when he saw the whooping cranes, she said, and every time their sprawling wings unfolded from their bodies and the three birds lifted off, he'd almost lose his breath. "I was glad," Gibbs told me, "that he still had enough to know that these beautiful birds had visited us."

Fifty-three million Americans feed birds outside their houses. But the First Family had chosen Clarice Gibbs, and so here I was. Her story was too idiosyncratic—too sad and beautiful at the same time—

to reduce to some clear and prescriptive moral. Maybe every story about people and wild animals in America is.

Instead, it illustrated only a universal paradox: how wildness fulfills certain human needs and is also trampled by them; how easily we can wind up short-circuiting and celebrating it at the same time. Here was a fundamental tension I'd noticed everywhere, in every era, crunched into the space of one backyard. All Gibbs had really done was appreciate our planet's flashes of wonder so intensely that she couldn't bear to watch them go away. It was a familiar longing, the same one that sends other people into careers in conservation.

When we were done talking, she walked me to her front door and thanked me. I was exhausted and needed to drive four hours north to a place in the Panhandle called Jefferson County to reconnect with Operation Migration. Two months after leaving Necedah, they had finally crossed into Florida.

As I made for my car, she called after me to stay safe. "There's lots of nitwits driving out there," Gibbs said.

15.

BACKPACKS FULL

OF ROCKS

Jefferson County, Florida, is a spare and bucolic spot east of Tallahassee. It bills itself as "the other Florida" and brims with weeping oaks and meandering roads. The county is named for America's third president and most zealous moose collector. Coincidentally, I read that, because of some fluke of geology along the Aucilla River, archeologists have dug up many prehistoric mammoth bones in Jefferson County.

Operation Migration had gotten the cranes to Jefferson three days earlier, setting down in a tremendous grass field miles from anything after another hot streak of flights. Brooke parked his trailer next to a pile of gravel and camped there alone, preferring solitude and the company of the cranes, while the rest of the crew found hookups for their RVs at a nearby KOA campground. All of a sudden, OM was threatening to polish off the migration, likely their last ever, within days.

In Jefferson County, Operation Migration's route forks. Ever since the storm surge killed the class of 2006 at the Chassahowitzka National Wildlife Refuge, the cranes have been divided between two refuges for the winter, to hedge against another disaster. Half the birds are taken to Chassahowitzka, but half are deposited farther north first, at the St. Marks National Wildlife Refuge on the Gulf Coast. It's a quick twenty-eight-mile hop to St. Marks from Jefferson County, and since the weather looked marginally better in that direction, the crew was staging to deliver those birds first. The refuge staff and the town of St. Marks were standing by to throw their annual arrival party, in which hundreds of Craniacs come out to witness the trikes and cranes fly in. A stage is set up. The crew gives speeches. It lends a feeling of finality to the months-long epic, even if the other half of the flock hasn't yet reached its home.

It was well after dark by the time I got to the KOA campground from Gibbs's house. I knocked on the door of Operation Migration's largest RV—the crew's collective living room—and found Joe and two other crew members, Walt Sturgeon and Jack Wrighter, watching *Antiques Roadshow* in a near-anesthetized trance. They'd been grounded in Jefferson County for two days already. The wind was hostile, and the temperatures at sunup had been so obscene—twenty degrees, seventeen degrees—that, elsewhere in Florida, manatees were huddling around coastal power plants, warming themselves in the outflowing plumes of hot water, and volunteers were kayaking into the Gulf of Mexico to ladle out sea turtles that were floating belly-up, stunned by the cold, then rushing them to triage centers and bundling them into blankets.

I'd expected that once OM got this close to their destination a levity or forgiving feeling of mutual accomplishment might set in. Instead, there was only the numbness of shift workers still a few hours from the end of a long day. On the television, a woman in San Antonio stood next to her set of Chinese porcelain paintings and told the appraiser, "It's called *The Four Seasons of Rice Planting*." You could see her practically pulsating, her hope that these heirlooms would be worth a fortune working hard to outstrip the shame of hoping for anything at all. The camera zoomed in on the joint of one bamboo frame. Walt, who has apparently done some woodworking, muttered from the RV couch, "The corners don't even match up."

Without looking away from the television, Joe and Jack both said, "Nope."

Just then, someone read out the next day's weather forecast from the RV's back room. What he said was impenetrably numeric to me—lots of knots and altitudes—but, given the reaction in the RV, which was no reaction, I took it as bad news. It wasn't. The forecast looked decent, but after seven states and sixty-five days on the road, no one wanted to believe it yet.

THE NEXT DAY, as it happened, was another squally down-day. After some waffles at the KOA, I picked up Brooke at his trailer and we drove to the St. Marks National Wildlife Refuge. The tracking team was reporting that six older whooping cranes had recently returned to the refuge for the winter—the first birds to finish their southward migration that year—and we were hoping to catch sight of them. It seemed as good an excuse for an outing as any.

The trip that Operation Migration had been waiting days to make in the air took less than an hour in my rental car. There was something immediately rejuvenating about St. Marks. Every migration stopover is purposefully remote, a cloistered dimple in the woods. But here the scenery was wide open and brand-new, seared with prismatic greens and blues. New kinds of birds—ibises and other shorebirds—flew in low lines between marshes.

This was a homecoming for Brooke—he spends every winter at the refuge—and at the visitors' center, a trio of women trickled out of the office to welcome him back. They'd printed out the online bios for that year's cranes and were eager to know which five of the ten were coming to live at their refuge, pressing Brooke for their "secret names." (Though Joe insists on referring to the birds by number only, Brooke finds it impossible to train them so closely and not give them nicknames. "Nobody has a kid named 'Number Thirteen,'" he told me.) Brooke smiled coyly and divulged only one secret name, "Rocky." Then, after some more chitchat, we were on our way.

We wound up on the beach at the foot of the refuge's old lighthouse and gazed for a while at the water as the sky turned overcast. I don't remember what I asked Brooke, but it must have touched a nerve, because he started, once again, lamenting the discord inside WCEP and the friction between him and the other trike pilots. The other day, he said, he was at a Walmart, making a run for cut-rate canned foods, when he turned and saw a little boy at the leg of his mother. The child was pointing at Brooke—he recognized him from OM's Web site—and gawped, his mouth hung open so wide that Brooke could see the socket of a missing baby tooth. It made Brooke feel small inside. "I feel like I'm scamming these people—at least a little bit," he said. By now, it's as though he and his coworkers have been turned into human polar bears, emblems that some people desperately want to rally around and that help them believe in good. We, on the ground, are imprinting on these costumed men, too, hoping to be shown what's possible with enough ingenuity and hard work.

Brooke seemed to understand this. But, he said, he also knows how uncivil the partnership has been lately behind the scenes, and how guilty he is of getting carried away with his resentments. And he knew that if he let even a speck of that cynicism come across, standing in the aisle at Walmart, it might shred something priceless inside that little kid.

All fall, I'd been stunned by how open everyone working on the whooping crane reintroduction had been with me—and especially the people of Operation Migration, whom, as the migration proceeded, it was clear I'd caught in a moment of great rawness and uncertainty. Now I leveled with Brooke. I told him that the coarseness I'd encountered in the partnership was making me lose hope a little, too. But how could he feel that same disillusionment and still throw himself into the sky every morning to push the birds forward?

"I'm just fucking stupid," he said. He was almost shouting now. "Believe me! At the end of the day, I want to call up my son and beg forgiveness. Because it was all an accident. I wanted to get my son involved in birds, because it seemed like a good thing to do, and I didn't want to be a shitty father. And what did I do? I ran away to be a bird guy and wound up being a shitty father. Now he's in college and he doesn't even answer the phone half the time I call.

"I didn't give a shit about birds at the beginning," he went on. "I'm not an altruistic person. I'm not even really a nice guy." But the whooping crane project had materialized in front of him, and it was just as though he were driving down a highway, saw someone broken down on the shoulder, and knew he had a jack in his trunk. A very simple thought rose up in his mind: "I can do this. This is something I can do." Now he was determined to keep at it until the thing was fixed.

"The way I look at it," Brooke said, "is that everybody, every one one of us, has this big sack around our neck—a backpack full of fucking rocks. Being evil, being bad, and being greedy—all the human shit, all of our frailties, everything that makes us human— they're all packed in a great big bag on our backs. We're born with it. For millions of years, humans have been loading that pack. So, unless you're Jesus or Mother Teresa, you've got this pack to carry. And we carry it into WCEP meetings. We carry it on the flight with the birds. We carry it into the motor homes at night. We carry it into our relationships—into parenthood. Always, always: you're carrying this fucking pack."

The whooping crane restoration, Brooke went on, is like a long swimming pool that everyone has to swim across together. "There's not many people that are going to swim with the fucking pack on their back full of rocks!" he said. "But a lot of the people in this project make that swim. And even though they're carrying all this negative shit, they're down in there: splashing, splashing, trying to get to the other side.

"I *know* I can sound negative. Everyone's always telling me, 'Oh, you're so negative.' I get out of those WCEP meetings and I'm just looking for a bridge to jump off of. But what I have to remind myself is that the people in that room were trying—they're just all carrying their packs. We're never going to shed that baggage, but at least we can carry it in the right direction."

The dysfunction didn't surprise him, in other words. It was infuriating, yes, but not something to be defeated by. He was staring at the same absurdity that Joan McIntyre saw, the way people's good intentions can so easily curdle when combined.

But he wasn't about to give in the way she did, to tender his resignation with bouquets of daffodils. There's an inevitable entropy in people—a wildness inside us—that tugs at the threadwork of everything we do. Even a job that looks so idealistic and decent can get pocked with misunderstandings, egos, and competition. But that's what every human enterprise is like. That's what every *ecosystem* is like. To call it unfair is a cop-out. It's the natural order that our species fights to transcend.

"You can't take humanity out of the equation," Brooke told me. "Humanity caused the problem to begin with, and so it's very hard for humanity to *solve* the problem. Because it's humanity! You know what

I mean? We bring to the table all the same crap that was brought to the table to create the nightmare in the first place!"

Maybe this sounds deflating—too jaded and profanity-laced to lay out for a kid in Walmart, or, for that matter, for a high-schooler looking into a polar bear's eyes on Buggy One. But I swear, on the beach at St. Marks, it felt like the most reassuring thing I'd heard any conservationist say. This is all there is, and all there ever could be: achingly imperfect people, working to achieve something more moral than they are. "It's not a bird project," Brooke said. "It's a people project. The birds are an excuse for doing something good."

EARLY THE NEXT MORNING, with a plum-red sun still concentrated at the horizon, I was scrunched into the backseat of a Cessna, looking down on Jefferson County's quilt of cattle pastures, farmhouses, and radio towers.

Piloting the plane were Operation Migration's "top cover" pilots, Jack Wrighter and John Cooper—two retired commercial airline pilots who, on migration days, soar a thousand feet over the trikes and cranes, working the radio to clear any air traffic ahead, and keeping an eye out for wayward birds.

This morning, the five whooping cranes bound for St. Marks were finally going to fly the eighteenth and final leg of their migration. While Joe Duff's trike soared high and kept its distance, Brooke circled toward the lightbulb-shaped field where the cranes were penned. "Okay," I heard him say over the radio. "Let 'em out."

The door opened. The cranes swaggered out of the pen, rather than surged, then gathered into the air lackadaisically as Brooke made a low pass to pick them up. They didn't form a line at his wing, but vibrated slackly behind his trike as they climbed—five shapes, pumping like uncoordinated pistons. There was a protracted power struggle—a "crane rodeo," as the pilots call it. But within twenty minutes, Brooke had sorted the five birds behind his left wing, and the migration settled in at twelve hundred feet. Soon we were crossing the Wacissa River en route to St. Marks.

Joe was drifting five hundred feet above Brooke. We in the Cessna were five hundred feet above Joe, flying in broad circles to keep from creeping too far past the convoy. Ahead, I could see the tidy rows of pine forests bisected by roads, and the long coastal highway, mirrored by a parallel highway of high-tension wires. Beyond that, a silver power plant rose out of the flat earth, glinting and disgorging dull white smoke. Then, at the Gulf Coast, everything faded to blue. Jack, the Cessna pilot, announced that we were thirty-one minutes away from the wildlife refuge.

At that moment, six hundred people and a couple of local television crews were waiting on a long, carpetlike lawn at the edge of the town of St. Marks, just across an emerald estuary from the refuge. Among them, I'd later learn, was a man from Ontario, Canada, who, a few days earlier, had read online that the migration was closing in on its endpoint and asked his girlfriend, "Want to go see the whooping cranes fly in?" An hour later, they were in their van, driving to St. Marks. There were two sisters, ages nine and twelve, whose parents had offered them each $25 to go to school that day, but who turned down the money so they could fake the flu and come collect autographs from Operation Migration's crew. There was an elderly woman who, inexplicably, had brought a used space pen—the kind that writes upside down—as a present for Brooke. And there were two doughy, jovial young guys—volunteers at the refuge—who'd sneaked into the observation tower next to the pen where the cranes were being delivered and, given that it was now the middle of December, nailed up five Christmas stockings, embroidered with each of the whooping cranes' numbers, and a wooden sign that read HOME SWEET HOME.

Thirty-one minutes from now, they would all look up and watch Brooke's trike pass directly overhead at five hundred feet, low enough so that, when he banked left to steer the cranes over for a second showing, everyone on the ground could see the broad white arm of his costume waving to them.

Soon we were twenty-four minutes away. Then eighteen minutes out. The whooping cranes hadn't so much as twitched from their places behind Brooke. There was nothing to do. In the Cessna, Jack reached into the backseat beside me and pulled a granola bar from the pocket of his coat.

VERY LATE THE PREVIOUS NIGHT, I'd run into Joe at the KOA campground's game room. I was looking for Internet access. That winter, when I wasn't on migration, I was home touring preschool after preschool with my wife, and now I'd promised that I would bang out all the applications before I came home from Florida.

It was dark out, and through the game room window, behind a nativity scene of porcelain miniatures, I could see Joe, sitting in a folding chair in an Operation Migration sweatshirt, one arm drooped across the exercise machine beside him. He had a load of laundry going in the coin-op washer outside and was waiting to dry it. His head was winched up—uncomfortably, it seemed—so he could stare at the television mounted from the ceiling. The only light in the game room came from a Christmas tree in the corner and the flickering of *Two and a Half Men*. I stopped for a second on the path outside, because I wanted to remember the image. It was a portrait of the twenty-first-century conservationist on the road—a man without the luxury of believing that anything would be simple, or of any expectation of closure, but still trying.

In the end, the embittered atmosphere around the whooping crane reintroduction had only made me admire everyone in the partnership more. Because, even as they thwarted themselves, everyone was still trying. There was no getting around that fact, and so much dignity in that persistence. I had never seen courageousness that looked quite like it.

It was heroism in the Sisyphean sense, just as all conservation may be, since implicit in that work now is the impossibility of its ever being finished. It requires rallying the will to build something that the future is likely to erase. Even Joe, in one of his darker moods, had admitted to me, "There's no hope for whooping cranes in the long term." Climate change, for one thing, will put much of the marshland we've protected for the species underwater. Joe seemed acutely conscious of how much climate change might upset or wreck. And he worried about the plastic infusing the world's oceans, too, and overpopulation—the whole litany of environmental terrors. "I've been involved in this for so long, I'm to the point now that I wish I hadn't had a daughter," he confessed. "I worry my daughter's going to have a shit life. I really do. That's scary. I mean, we've left a mess. I think the world is facing a major crisis, and nobody cares. And I don't know what to do, do you? But I feel like I'm doing something useful. I'm doing *something,* and I don't know if it's going to work out or be worth anything, but at least we're trying."

I didn't have any better answers. After everything I'd seen, I had no solution for fixing a broken world. But, then again, that's only one of the problems we're facing. Another is just figuring out how we are supposed to live in a broken world. And, I realized, more than anything having to do with wild animals or biodiversity or the environment, *that* was the problem that I'd been wanting to give my daughter a head start in solving for herself.

BACK IN THE CESSNA, a voice cut in on the radio: "You guys still on course?" It was Liz, Operation Migration's woman on the ground in St. Marks. She was standing on the stage at the party, killing time as best she could, reminding the crowd of Craniacs to buy their Operation Migration T-shirts and videos and whooping crane Christmas ornaments. People were desperate for an update. I could picture them assembled on the lawn, blowing into their hands or resting them on their sons' and daughters' shoulders, all turned to face in the same direction, searching the sky for movement.

I suddenly realized that none of us in the air had said a word in a long, long while; the radio had been silent. I don't think Brooke had spoken at all. I'd spent most of the flight staring out my window at him in his trike, a thousand feet down, with the five whooping cranes fastened to his wing—the six of them in white, tied together as one object and hanging over Florida like a mobile above a crib.

It's what I remember most—not the arrival, but the long, uncertain lull beforehand.

"We're eight miles out," Joe finally answered.

"We're looking for you," said Liz.

THE MAN WHO
CARRIED FISH

started out from home early one morning, long after the whooping crane migration was finished, closing the car door softly, and rolling away from the curb under Isla's bedroom with the headlights off, so I wouldn't wake her. Seven hours later, having twisted through the snow-caked Sierra and wound south, I reached Phil Pister's house in a small desert town called Bishop, California.

Pister had celebrated his eighty-third birthday two weeks earlier. His skin was bronzed and rough-looking and pooled off his hands. I'd come to ask him to tell me a story—a story I'd heard in the course of my research a couple of times, but only secondhand and in the broadest strokes possible, as though it were a myth from outside of time. It was the story of one man—Phil Pister—who was tested in the desert. People called it the "Species in a Bucket" story.

It was August 18, 1969, Pister began. He was forty, and already fifteen years into his career as a fish biologist with the California Department of Fish and Game, here in Bishop. One afternoon, a young guy who worked for him named Bob Brown swept through his office door. The kid looked on edge. "And he says, 'Phil, we've got to get out to Fish Slough, because if we don't, that pond is going to dry up!'"

Fish Slough is a narrow, marshy depression in a valley of pale green scrub north of town. Water burbles out of the sand from a handful of springs at the north end, then flows south before heading for the Owens River. Much of the land in and around Bishop is owned by the city of Los Angeles. For almost a century, the city has reengineered the landscape into a giant funnel, diverting water to its residents more than two hundred miles away. Water has also been pumped out locally. Gradually, a rambling network of pools tightened into a thin spindle of water. And that particular summer, during a long heat wave, the high grasses were sucking the slough dry without anyone's noticing. Except Bob Brown, the kid in Pister's doorway. "What Bob didn't say when he came in," Pister explained, "but what was very obvious, is that if that pond dries up a species is gone."

The Owens pupfish lives in the pools at Fish Slough and nowhere else in the world. It's two inches long and otherwise so mundane-looking that it escaped my powers of description, no matter how intently I stared at the photo of one on Pister's wall and tried to jot down something meaningful about it in my notebook. It is beige.

One biologist told me that, before European settlement, the Owens pupfish lived in Fish Slough "by the zillions." But by the 1940s, with the pools in Fish Slough shriveling, the species had been presumed extinct. It was rediscovered by two visiting scientists in 1964. Pister was in the field with them, and, he told me, seeing a small riot of these ghosts still zipping around in the water, he experienced something comparable to a religious enlightenment. All of a sudden, the work he did for the state government seemed worthless. "Not only did I drop everything right then, Jon," Pister said, "but I never picked it up again."

There are roughly one thousand lakes in the region around Bishop, and Pister's job at the time was to fill them, and keep them filled, with the species of fish that fishermen wanted to catch. America has been stocking its waterways with nonnative fish at least informally for more than a hundred years, just as game wardens might introduce deer, elk, partridge, and other animals for hunters. (Even John Muir, the prophetic champion of raw wilderness in the Sierras, advocated fish stocking. Good fishing opportunities, Muir argued, would draw Americans into the mountains, where they could be rejuvenated. Trout were "bait for catching men, for the saving of both body and soul.") When Pister was on the job, fish stocking was a multimillion-dollar state-run operation. (Like many states, California still does a fair amount of it.) Trucks dumped trout and bass, mass-produced in hatcheries, into lakes and streams near the roadside. Mules carried them into the backcountry. At the most remote lakes, airplanes flew overhead and opened their hatches, and fish came raining down. In a way, the entire undertaking was a mirror image of the monumental salmon conservation project that the fish counters I'd met are a part of at Bonneville Dam. Back then, Pister was keeping rivers full of the fish that America thought were most delicious. Now there's a movement to keep them filled with the fish we think are most ecologically appropriate. Either way, though, you could argue, we are engineering nature to satisfy our taste in fish.

Seeing the Owens pupfish still alive in 1964, Pister suddenly flipped from one mind-set to the other. He realized he'd been devoting his life to supplying anglers from Los Angeles with coolers full of trout. Meanwhile, the region's native fish were quietly vanishing—largely into the stomachs of the bigger, hungrier fish that Pister was setting loose in their territory. And so he started working behind his bosses' backs to restore Fish Slough, trying to give the newly resurrected Owens pupfish a fighting chance to rebound there. Soon he was laboring for the benefit of other native fishes, too. That is, Pister was spending enormous amounts of his own time and money undoing exactly the sort of "improvements" that he was engineering at his day job.

By 1969, when Bob Brown came to Pister's office door, there were about eight hundred pupfish living in a single small pool at Fish Slough. This was the pond that Brown was warning him about, the one going dry. Pister grabbed Brown and another coworker and raced to the site. There was still water left in the pond, but it was hot and virtually stagnant, bereft of oxygen. Many fish were dead. The men netted as many of the living ones as they could and transplanted them into "live cages"—small cloth-sided boxes that allow water to circulate through, but keep the fish contained. Then they put the live cages closer to the mouth of the spring, where the water was flowing more rapidly. The ordeal was over, it seemed. It was dark now, Pister told me. "I told my buddies to go get something to eat."

Pister watched them drive away and started cleaning up. Before he headed home, he decided to take one last look. The situation had changed. Many of the pupfish in the live cages were now swimming belly-up at the surface, knocking into the thick, drifting traffic of those that were already dead. "I could see the bottom line on this. None of them were going to be alive much longer," he told me. "And that is when I went across the marsh, holding those two buckets."

He had two five-gallon buckets in his truck—the big white kind that they make pickles in, he explained. He filled them with water and began scooping up the fish that were still alive and putting them into the buckets. Then he lifted the handles and started to walk.

It was two hundred yards back to his truck. Pister had taken a few steps when the harrowing thought occurred to him that he was carrying the totality of a species in his two hands. He scanned the ground for snares of barbed wire and other obstacles, trying not to trip. Then he loaded the buckets onto his truck bed, sat in the driver's seat, and turned the key.

He'd made a snap decision to move the fish to another spring at the opposite end of Fish Slough, where he knew the water was cooler and deeper. The two springs are about a half-mile apart and connected by a doglegging dirt road that is packed with rocks; it is not a smooth ride. I asked again and again, but Pister couldn't remember what was going through his mind that night as he crept through the dark in his pickup, the two buckets of fish standing side by side on the truck bed behind him, joggling and splashing as he drove. I could almost see him: this minor Noah, moving the animals through an imperceptible flood. But there was a matter-of-factness to Pister's telling: he was confronted with a small, obvious crisis and saw a way to solve it, so he did. When he finished telling his story, Pister just looked at me and said, "It's something you do, you know?"

In 1931, near the end of his life, William Temple Hornaday began his last book by slinging insults at America's most prominent conservationists. He had given up on groups like the Audubon Society; they seemed to him cowardly and ineffective. Conservationists were not saving anything from destruction, he wrote, but only "humbly meandering along behind the firing lines, picking up the cripples."

Hornaday called the book *Thirty Years War for Wild Life: Gains and Losses in the Thankless Task*. It was both a memoir of his many campaigns and a continuation of the same spittle-ejecting invective he'd been writing for most of his life. By then, he'd become insufferably morose. In the book's final section, called "The Curtain," he wrote, "Regarding the future of the wild life of North America and the world at large, the author of this volume is a calamity-howler and a pessimist of the deepest dye." He had tried to "inject humanity and courage into the hearts of men," to awaken them to the value of their nation's wildlife, but now believed it couldn't be done. Maybe, he wrote, the only hope for preserving animals now was to control the appetites of Americans "despotically" with "the power of a Mussolini."

"Think it over," Hornaday wrote. "But, meanwhile, prepare for the Worst."

It's now been forty years since America passed the modern Endangered Species Act, committing itself to a brand of ecological idealism that really isn't so different from what drove Hornaday. It's pretty simple stuff, rooted in the same lessons that Isla is now learning at preschool: Be considerate of others. Don't take more than your share. Clean up your mess.

Understandably, beliefs like those will start to feel confused when the children who have internalized them start applying them to a larger, more complicated world. Even the situation of the Owens pupfish isn't as simple as the "Species in a Bucket" story makes it sound. I'd imagined Pister, having reached the end of the road, tenderly tipping the buckets into the new spring and watching the pupfish scatter. But in fact first he had to poison out all the other, predatory fish there. Then, later that summer, he built a gravel dam to shore up that new sanctuary—and on and on the management went, and still continues today.

When I visited Fish Slough recently, a retired government biologist named Terry Russi told me that the pupfish are "really no better off than they were forty-five years ago," when Pister moved them. The species lives in only four tiny pools, one about the size of the truck that Pister rescued it in, and right-wing residents of Bishop, with their ideological disdain of government projects, have occasionally tried to sabotage the operation by putting bass in the slough to gobble up the endangered fish and end the whole affair. By now, the government doesn't even disclose the locations of some Owens pupfish habitats, and there are no signs about the fish at Fish Slough. Instead, the spring where Pister brought his buckets that night is fenced with barbed wire, attended by a $17,000 instrument to measure its flow, and block-aded at one end by a fish-proof barrier to keep out predatory fish and keep the pupfish inside. The scene was part witness-protection pro-gram, part hospice. Standing at the edge of the pond, Russi explained that if vegetation is allowed to build up on the barrier's stainless-steel grate, the way scirpus and bulrush reeds had collected there now, the barrier won't work—predators can swim right over it. And so, mid-sentence, Russi suddenly dropped to all fours and started pulling the reeds out.

This is what a lot of America's war for wildlife looks like forty years into its current phase, ten years longer than Hornaday had been fight-ing his own war when he finally seemed unable to stomach it any longer. In fact, over the last ten years, support for all kinds of environ-mental causes has sagged in our country. Conservationists have been driven to take more absolute and truculent stands, because they be-lieve that even the smallest defeat will erode the bigger principles. They've opposed solar panel arrays and wind farms—green energy projects that would, in theory, help all species—for the sake of the desert tortoises or imperiled birds that live nearby. Meanwhile, on the other side, fuses have started blowing. During the time I was writing this book, I kept noticing headlines about horrendous explosions of spite toward protected species: the corn and soybean farmer in Min-nesota, for example, who in 2011 became so irritated by American white pelicans trampling his crop that he eventually snapped and smashed thousands of their eggs and stomped young chicks to death. Since 2007, eleven whooping cranes have been found shot dead in the wild. One was the matriarch of the First Family, and in that case, even though government investigators found the shooters and got a conviction, an unimpressed judge in Indiana sent the men away with a $1 fine.

This is why, when I'd first heard the "Species in a Bucket" story, I assumed it wasn't true. It sounded too pat, like just the sort of fable that American conservationists would need to tell themselves at a time when their work has become mercilessly convoluted and drawn out. The simplicity of what happened at Fish Slough in 1969 seemed to absolve everyone and bring the entire issue back into comprehensible terms—a basic place of agreement from which we might start over. There are people who wouldn't have gone through the trouble to move those fish, had they been in Phil Pister's place. But if you strip away the politics, the money, the philosophical arguments, and the appeals to moral responsibility—if you just see the man and the fish— I'd like to believe that almost everyone can appreciate why he did. Even Isla understood it. "Because fish need water," she explained, when I got back from Bishop the next evening and told her the story.

"WHAT A GREAT STORY," Brooke Pennypacker said.

It was September, a year after the whooping crane migration I'd followed to Florida began, and Brooke and I were having dinner at a roadhouse deep in central Wisconsin. I'd just repeated the legend of Fish Slough for him, and Brooke, who saw life as nothing but a chain of stories, was now visibly turning over the implications of this one in his mind.

Operation Migration had survived—at least for another year. That January, the Whooping Crane Eastern Partnership finally found a new site from which to launch an ultralight migration, now that Necedah was off the table: a remote state wildlife preserve called White River Marsh. The Wisconsin Department of Natural Resources hustled to get all the permits hammered into place only two weeks before Operation Migration needed to start training cranes. Then a gaggle of Craniacs arrived to help. A pen needed to be built. A pond had to be dug. Trees had to be felled to clear a runway for the trikes. They worked in a downpour, like neighbors raising a barn.

The week I visited White River Marsh, Brooke was the only pilot on-site, handling the flight training with two interns. I found the crew's familiar squadron of RVs parked in overgrown grass around a dilapidated farmhouse. The pig farmer who'd owned the property had died, and the state of Wisconsin planned to demolish the house and fold the land into the wildlife preserve. It seemed possible, though, that some of the structures might crumple to the ground first on their own. Kittens skittered near the squalid barn like street kids in a Dickens novel. What I mostly noticed when I got out of my car, though, was the smell. "There was a pile of pigs over there until a couple of days ago," Brooke explained.

I'd expected to find Brooke wallowing in exile. After all, the fate of Operation Migration was hardly less uncertain than when I'd last seen them. And the migration the crew was about to start that fall would turn out to be their most exasperating yet. They would lose one bird mid-flight before even leaving Wisconsin—not have it die, but physically lose it—and later be grounded briefly by the Federal Aviation Administration because of some oblique quibble the agency had with OM's nonprofit status. By Christmas, the crew would only reach Alabama, and be stranded there so long because of weather that, when they tried to fly again, the birds—likely assuming they'd reached their winter home, and losing their physiological urge to migrate—would refuse to follow. The cranes would be boxed, driven to a nearby wildlife refuge, and left to spend the winter there instead. For the first time, in other words, OM would fail to make it to Florida.

Still, right now at least, Brooke was upbeat. When I asked if it was tough to be shunted onto this isolated pig farm, he told me that he appreciates the solitude, actually, and that the public library in the nearby town is much better than the one in Necedah. "Last year at the refuge was not a comfortable situation," he said. "It was a slow tightening of the rack." He was, in fact, much happier and looser than when I'd last seen him, as if he'd somehow wound back the mileage accumulated the previous fall and restored the good humor that had gotten sapped out of him. He was back in his element—which is to say, he was pushing as hard as he could, in all directions, against many small and immediate frustrations: the birds that were slow to follow his trike during training; the pig stench. "The little picture," Brooke said, "is a whole lot easier to deal with than the big picture. That's for sure."

You could argue that this is the crux of a terrible problem. In the end, I can't say I'm optimistic about the future of wildlife. The stories of the polar bear, the butterfly, and the whooping crane had, at times, even lowered my confidence in our ability to see the problem clearly. There's a fluidity to nature that's not easy to recognize or accept, and climate change will only accelerate and distort such changes. There's also a fluidity to how we feel about nature—the way our baselines subjectively reset and will keep resetting far into the future, while, in the background, the empirical damage piles up.

These are destabilizing thoughts. I still don't know what to do with them. But neither does anyone else, it seems, and so their weight has a way of compressing conservation down into a nearsighted exercise— one that can be pursued only by focusing on the little picture of the present and by blocking out the yawning uncertainty that moment is adrift in.

Then again, it was people's capacity to focus on that little picture that I found so invigorating everywhere I went—and to keep refocus- ing on it; having reached a place of devastating pessimism, to return somehow to a place of, not optimism exactly, but at least relative equanimity. Maybe what conservation tries is sometimes misguided or futile. But there's something deep and blameless in the trying itself—a spark we can feel defined by as humans and should point out to our kids.

Invariably, the many battle-scarred old conservationists I met told me they failed. But, really, having been thwarted, they usually just refocused that same moral energy in different ways. For all his saltiness, Jerry Powell, the lepidopterist, still Xeroxed and mailed a copy of his Antioch Dunes study to every Fish and Wildlife employee who asked for one. Joana Varawa told me she was now striving to act compassionately toward those immediately around her on Lanai rather than gather up the compassion of all humanity to save the whales. Even Rudi Mattoni, who told me he'd given up—who used those exact words—had definitely *not* given up, but was instead settling into his corner of the jungle with a net and a notebook, to preserve the memory of every last insect he could.

If they'd all given up, why did they spend so much time talking to me so I could write their stories down? And how was it that, during those conversations, they could access such ardent feelings of frustration all over again, if they weren't still trying to unscramble the same question that I'd apparently gotten stuck on, too: how, on earth, should we human animals live?

I'd been picking apart the stories we tell about wildlife, hoping to find a firm conclusion, or even some new and useful vision for our shared future together. But I never came close. America rewrites those stories so erratically over time, and so impulsively, that few of them feel convincing in the end. Instead, I'm convinced by the stories that we use wild animals to tell about ourselves. The best of us are cursed with caring, with a bungling and undying determination to protect whatever looks like beauty, even if our vision is blurry. People kept warning me that Isla's generation will blame us for losing so much of that beauty. But whatever: it's inevitable, and I'm trying to make my peace with it. It's comforting that they'll still imagine better, and it will occur to them to be angry.

As THE FOG CLEARED at White River Marsh the next morning, I stood in a blind with a small group of Craniacs and watched Brooke train the new class of birds. It seemed to go very badly. ("I think they've still got a lot of learning to do," an old lady next to me whispered.) Afterward, Brooke rushed out to a cement-slab hangar that OM was renting, to meet two inspectors from the Federal Aviation Administration. The FAA had called out of the blue, wanting to inspect their trikes.

Given the crew's trouble with the agency later that fall, the inspection feels foreboding in retrospect. Brooke may have half understood that at the time, but while the two men meandered around each aircraft, ticking boxes on their clipboards, he merely did what he always did: he talked. He explained the project to the inspectors—the costumes, the imprinting, the long push to Florida—and, before long, seemed to be winding his way back to a conversation that he and I had had at dinner the night before.

He'd told me about one morning that spring, when he was still at the St. Marks National Wildlife Refuge in Florida. Two of the cranes I'd watched Operation Migration deliver to the refuge at Christmas had already left on their northbound migration back to Necedah. Brooke was waiting for the last three to decide to depart. At sunup one day, after checking the birds and doing his usual battery of chores, Brooke—playing a hunch—settled into the blind next to the pen to watch. (Half of the pen has no top. The birds come and go as they please all winter, typically exploring the refuge by day and flying back in to roost at night.) Sure enough, the three birds soon issued a long, blaring call and launched. They coiled higher into the blue. Brooke watched them vanish. On his receiver, he heard the beeps of their tracking bands get softer and softer as the cranes soared out of range. Then, finally, he couldn't hear anything anymore. They were on their way.

Immediately, Brooke told me, he started packing up. He hurried around the blind, taking the maps off the walls, and walked into the pen to unhook the feeders. "Then I say to myself, 'Slow down a minute.'" This was it, after all—the moment everyone worked for. For the first time since the cranes had hatched at Patuxent, they were on their own. They were wild. Soon Brooke would drive to Patuxent and start the process all over again, with new chicks. But the brief sliver of every year between the end of one cycle and the beginning of the next had started the second those three birds disappeared in the sky.

Brooke walked into the shed next to the pen and sat down among the leftover bags of crane chow. It used to be that, after dropping the whooping cranes at the refuge in Florida, he and the other pilots would bounce out of their cockpits and start high-fiving each other and hugging. But in the past few years, it hadn't even occurred to them to celebrate; they touched down and, without a word, began briskly dismantling the wings of their airplanes and bagging their gear. "There are many triumphs through this whole experience," Brooke told me, "but there isn't a lot of rejoicing anymore."

So, sitting in the feed shed, Brooke decided to share that moment. He called his girlfriend, who used to work for Operation Migration. He called Joe. Then he started calling Craniacs, stopover hosts—many different kinds of people, all of whom had a hand in the project and who, Brooke realized, deserved to feel the same momentary sense of accomplishment—the *illusion* of closure, at least—that he was now forcing himself to feel. He must have called a dozen different people, he told me. "And if the battery in my phone would have lasted, I could have stayed in there calling people for four or five days, when you think of all the people that have helped this project go forward."

It was this collaborative aspect to the reintroduction that he was explaining now in the hangar, though the FAA inspectors seemed almost not to hear him as they knelt and prodded one of the trike's propeller guards. Soon even I was only half listening, packing my backpack on the seat of my car, waiting for a free moment to say a last goodbye and head home.

It was almost as if Brooke was standing in the hangar telling stories to himself, marveling about how, a few months out of every year, people welcome these strangers onto their land, open their homes, and keep them well fed, all for the sake of a wild animal they'll probably never see—and how unusual that is, if you take a moment and think about it.

"These birds," I heard him say, "they've got a key that just unlocks the goodness of people."

ACKNOWLEDGMENTS

This book exists only because so many many knowledgeable, compelling, and decent people—those I've mentioned and others I have not—were willing to talk with me about the work they do and the causes they care about, or even invite me into their worlds to see for myself. For that, I thank them.

In 2009, I wrote a story about Operation Migration for the *New York Times Magazine* that, in retrospect, birthed this book. I'm grateful to the many gifted editors I've worked with at the magazine—Jamie Ryerson, Paul Tough, Sheila Glaser, and Hugo Lindgren—for making me a better reporter and writer; to ace fact-checker Lia Miller; and especially to Alex Star and Gerry Marzorati for first letting me through the door.

Thanks to Doug McGray, Evan Ratliff, and the rest of the *Pop-Up Magazine* family in San Francisco for giving me a chance to try out my Billy Possum material on stage, and to Roman Mars of *99% Invisible* for putting it on the radio.

I first visited Churchill in 2005, five years before the trip I chronicle here, while a student at the University of California, Berkeley, Graduate School of Journalism, for a project on climate change called "Early Signs." Thanks to Sandy Tolan and Orville Schell for making that trip possible and to *Salon* and Public Radio International's *Living on Earth* for giving our reporting a home. Thanks also to Nick Miroff, my friend and collaborator on that project, with whom I first learned to see the stories unfolding there.

Thanks to Michael Pollan, for offering advice and a critical confidence boost at the beginning of this project; to Laurel Braitman, Jennifer Kahn,

ACKNOWLEDGMENTS

and Nick Miroff for reading parts of the manuscript at different times; to Chris Colin for reading parts of the manuscript all the time; and to Jack Hitt, for all of the above and more: "If you know him, you know why."

I'm grateful to my agent, David McCormick; my dogged fact-checkers, Timothy Leslie and Hamed Aleaziz; and to Lindsay Whalen, Terry Zaroff-Evans, and others behind the curtain at The Penguin Press, all of whom improved the manuscript, and my mental health, in many subtle but important ways. And I'm especially grateful to my editor Ann Godoff, whose trust and guidance was a gift.

Finally: thanks to Wandee, who I can never thank enough, and to Isla, who has a loose tooth and feels strongly that it should be mentioned somewhere in this book.

NOTES

INTRODUCTION: THE WOMAN
WHO COUNTED FISH

The projection that half of the world's species will go extinct by the end of this century gets thrown around a lot by conservationists and journalists. Thanks to Bradley Cardinale at the University of Michigan for helping me understand the research that prediction is based on and two key scientific papers in particular: David U. Hooper et al., "A Global Synthesis Reveals Biodiversity Loss as a Major Driver of Ecosystem Change," *Nature* 486 (2012) and Anthony D. Barnosky et al., "Has the Earth's Sixth Mass Extinction Already Arrived?" *Nature* 471 (2011).

I first read about training condors to avoid power lines in David S. Wilcove's *The Condor's Shadow: The Loss and Recovery of Wildlife in America* (New York: Random House, 2000), 195. And I first read about human-assisted salamander migrations in his book *No Way Home: The Decline of the World's Great Animal Migrations* (Washington, DC: Island Press, 2008). More important, conversations with David were a great help throughout this project; I feel lucky to have had his ear.

Ferret plague vaccines are detailed in "Ouch! Taking a Shot at Plague," a press release from the U.S. Geological Survey, July 16, 2008. (The agency has since developed an edible, "ouchless" vaccine.) "Spotted Owls Face New Threat," broadcast on the KQED radio show *Quest*, August 29, 2011, covers the management of spotted and barred owls in California, and the monitoring of pygmy rabbits is discussed in Eveline S. Larrucea and Peter F. Brussard, "Habitat Selection and Current Distribution of the Pygmy Rabbit in Nevada and California, USA," *Journal of Mammology* 89 (2008) and "Military Drones Spy on Pygmy Rabbits," by Dave Wilkins, *Capital Press*, July 8, 2011. I learned about the protocol for helping Alabama sea turtle hatchlings by spending a week with the volunteer "turtle people" of Share the Beach while writing about their role in a dramatic turtle rescue after the BP oil spill for the *New York Times Magazine* ("Night of the Turtle People," October 1, 2010).

I spent a lot of time studying efforts to protect salmon on the Columbia River, but ultimately couldn't include much of what I learned. Thanks to Scott Clemans and John Rerecich of the U.S. Army Corps of Engineers; to Ann E. Stephenson and Janet Dalen of the Washington State Department of Fish and Wildlife's fish counting program; to Walt Dickhoff at NOAA; and to Conrad Mahnken, a Washington State Fish and Wildlife Commissioner.

I read about the resurgence of crocodiles at Florida Power and Light's Turkey Point nuclear generating station in *Hope for Animals and Their World: How Endangered Species Are Being Rescued from the Brink* by Jane Goodall, with Thayne Maynard and Gail Hudson (New York: Grand Central Publishing, 2009), 80–81. That's also where I discovered the peregrine falcon "copulation hat." I also consulted *Peregrine Falcon*: *Stories of the Blue Meanie* by Jim Enderson (Austin: University of Texas Press, 2005), 143–46.

The idea of "conservation reliance" was proposed in J. Michael Scott et al., "Recovery of Imperiled Species Under the Endangered Species Act: The Need for a New Approach," *Frontiers in Ecology and the Environment* 3 (2005). In a more recent paper, "Conservation-Reliant Species and the Future of Conservation," in *Conservation Letters* 3 (2010), Scott and his colleagues determined that 84 percent of the species listed under the Endangered Species Act are conservation-reliant. I interviewed Mike Scott about these papers when I was first getting interested in wildlife conservation, and it was largely because of the questions Mike raised that I was drawn to write more about the subject. I'm grateful to him.

Pam Aus, who shot the video of her cat, a fox, and an eagle on her back porch, appeared on the *Today* show on March 30, 2012. I'm quoting from a seventy-fifth anniversary edition of Henry Beston's *The Outermost House: A Year of Life on the Great Beach of Cape Cod* (New York: Macmillan, 2003), 25.

My understanding of early American wildlife is rooted in many sources, but two were key. In "'The Liberty of Killing a Deer': Histories of Wildlife Use and Political Ecology in Early America" (PhD diss., Northern Illinois University), Andrea L. Smalley writes of English explorers sweeping up fish with brooms. She also quotes a 1614 letter by John Smith bragging that anyone with "strength, sense and health" could thrive on the new continent. In *Nature's Ghosts: Confronting Extinction from the Age of Jefferson to the Age of Ecology* (Chicago: University of Chicago Press, 2009), Mark V. Barrow Jr. stresses how the abundance of wildlife figured into the self-image of early Americans.

The estimate of thirty million buffalo I cite is included in James H. Shaw, "How Many Bison Originally Populated Western Rangelands," *Rangelands* 17 (1995): 149. William Temple Hornaday argued that protecting buffalo around Yellowstone was "in the interest of public decency, and for the protection of the reputation of American citizenship," in *Our Vanishing Wildlife: Its Extermination and Preservation* (New York: Scribners, 1913), 91.

PART ONE: BEARS

1. MARTHA STEWART ON THE TUNDRA

It would have been nearly impossible to write about polar bears in Churchill, and impossible to write about them well, without the cooperation, openness, and hospitality of three groups of people: the staff of Polar Bears International, especially Robert Buchanan and Krista Wright; John Gunter and his crew at Frontiers North Adventure; and, of course, the many residents of Churchill who I met on my trip.

I learned about Churchill during the Cold War from the archives of the *Winnipeg Free Press* and by inviting myself on a tour of the Churchill

Northern Studies Centre, a nonprofit scientific research station on the site of the old rocket range, led by the Centre's director Michael Goodyear.

The 2007 government study I refer to is Steven C. Amstrup, Bruce G. Marcot, and David C. Douglas, "Forecasting the Range-wide Status of Polar Bears at Selected Times in the 21st Century," a U.S. Geological Survey Administrative Report. I read about maple syrup shortages in the USDA Forest Service's News Release No. 1022, "Climate Change May Impact Maple Syrup Production."

Conversations with Anthony Leiserowitz, director of the Yale Project on Climate Change Communication at Yale University, and Gavin A. Schmidt, a climatologist at the NASA Goddard Institute for Space Studies, informed my understanding of how Americans think about climate change. The poll from the Pew Research Center, "Little Change in Opinions about Global Warming," was released on October 27, 2010. Congressman John Boehner called the idea that carbon was harming the environment "almost comical" on an April 19, 2009, broadcast of *This Week with George Stephanopoulos.*

I spoke with Christopher Andrews, the director of the Steinhart Aquarium at the California Academy of Sciences and the academy's chief of public engagement, about why the museum was overhauling its climate change exhibit. (Chris also shared an internal audit of the exhibit's effectiveness.) And I learned about the polar bear "transition center" from Don Peterkin, Gordon Glover, and Douglas Ross at Assiniboine Park.

John Hadidian, director of Urban Wildlife programs at the Humane Society of the United States, first introduced me to the term "cultural carrying capacity."

2. AMERICAN INCOGNITUM

My account of the military's exit and the birth of the tourism industry relies on coverage of those eras in the *Winnipeg Free Press* but, primarily, on the conversations I had in Churchill with people who lived through them. Mike Macri, Paul Ratson, Mark Ingebrigtson, Dennis Compayre, Ed Bazlik, Claude Daudet, Myrtle Demeulles, Bob and Pat Penwarden, Don and Kyle

Walkoski, and Mayor Mike Spence were among the many good oral historians I encountered in town. Len Smith answered questions after I returned.

I learned about Ursula Böttcher from a short, March 30, 2012, obituary
in the *Telegraph* and from an April 9, 1980, *New York Times* article by Paul
L. Montgomery that describes how a New York City transit strike was
squashing attendance at the circus, leaving Böttcher to do her polar bear
show for a mostly empty arena. "It's a hard bread to eat, but you go on," she
told the paper.

I read several newspaper stories about the films *Polar Bear Alert* and *Blue
Water, White Death*, including a May 12, 1971, *New York Times* review of
the latter by Vincent Camby. Peter Benchley talks about his debt to that
film in the introduction to a later edition of *Jaws* (New York: Random
House, 1991), 1–2. Animal Planet's Plexiglas cube stunt is mentioned in
Chris Palmer, *Shooting in the Wild: An Insider's Account of Making Movies
in the Animal Kingdom* (San Francisco: Sierra Club Books, 2010), 148. Chris,
director of the Center for Environmental Filmmaking at American University, also stressed to me the influence that *Polar Bear Alert* and *Blue Water,
White Death* have had on the field.

My description of the Tommy Mutanen mauling and its aftermath is
drawn from both the *Winnipeg Free Press* and the firsthand accounts of
Mark Ingebrigtson, Mike Reimer, and Sandi Coleman, the television reporter I mention, among others.

I extrapolated the value of the polar bear tourism economy in Churchill
from "Evidence of the Socio-Economic Importance of Polar Bears for Canada," a 2011 report commissioned by Environment Canada, a government
agency, and prepared by ÉcoRessources Consultants—and specifically from
Figure 2: "Monetary Values Associated With Polar Bears in Canada, by
Value Category (Aggregate Amounts for Canada)." The tourism study I
mention is "Last-chance Tourism: The Boom, Doom, and Gloom of Visiting Vanishing Destinations," by Harvey Lemelin et al., *Current Issues in
Tourism* 13 (2010).

Hudson Bay's southern polar bear population has been studied for more than three decades by Ian Stirling, a retired adjunct professor in the department of biological sciences at the University of Alberta and research scientist emeritus with Environment Edmonton. I benefited from many studies he's authored and coauthored over the years, as well as his book *Polar Bears* (Ann Arbor: University of Michigan Press, 1999). A key source for understanding exactly how climate change will affect the town's bear population is "Polar Bears in a Warming Climate," *Integrative and Comparative Biology* 44 (2004), by Andrew E. Derocher, Nicholas J. Lunn, and Ian Stirling. An even more valuable resource for me was Andy Derocher himself. Not only did Andy talk me through the science with great patience and skill, he also looked over sections of this book before publication. I'm grateful to him for being so generous with his time, and also to Steven Amstrup, an equally gifted scientific explainer.

Robert F. "Rocky" Rockwell at the American Museum of Natural History talked with me about his goose research and the reactions it provoked. The scientific paper that established how futile it is for polar bears to chase geese is "The Significance of Supplemental Food to Polar Bears During the Ice-Free Period of Hudson Bay," *Canadian Journal of Zoology*, 63 (1985), by Nicholas J. Lunn and Ian Stirling. I first saw this research mentioned in Richard Ellis's encyclopedic book *On Thin Ice: The Changing World of the Polar Bear* (New York: Knopf, 2009), 95.

Early America's over-the-top enthusiasm for mammoths has attracted a number of perceptive historians. My writing on the subject relies primarily on books by three of them: Barrow, *Nature's Ghosts*; Paul Semonin, *American Monster: How the Nation's First Prehistoric Creature Became a Symbol of National Identity* (New York: New York University Press, 2000); and Lee Alan Dugatkin, *Mr. Jefferson and the Giant Moose: Natural History in Early America* (Chicago: University of Chicago Press, 2009). Dugatkin's book is also *the* indispensable resource for anyone interested in Thomas Jefferson's moose gambit. I also consulted Henry Fairfield Osborn, "Thomas Jefferson as Paleontologist," *Science* 82 (1935); Gilbert Chinard, "Eighteenth-Century Theories on America as a Human Habitat," *Proceedings of the American Philosophical Society* 91 (1947); and Ralph N. Miller, "American Nationalism as a

Theory of Nature," *The William and Mary Quarterly* 12 (1955). Especially useful for understanding Jefferson the man were Christopher Hitchens, *Thomas Jefferson: Author of America* (New York: HarperCollins, 2005) and Joseph Ellis, *American Sphinx: The Character of Thomas Jefferson* (New York: Random House, 1996).

Details about America's "mammoth fever" are covered in Semonin's *American Monster*, as well as in "The Cheese and the Words," by Jeffrey L. Pasley in Jeffrey L. Pasley and Andrew Whitmore Robertson, eds., *Beyond the Founders: New Approaches to the Political History of the Early American Republic* (Chapel Hill: University of North Carolina Press, 2004), 32–33. Semonin also describes a 1,230-pound "mammoth cheese," fabricated by a certain Baptist congregation in Massachusetts as a gift for President Jefferson in 1802. The cheese was six feet in diameter, made from the milk of nine hundred cows, and pressed on a cider press built specifically for the operation, engraved with the words "Rebellion to tyrants is obedience to God." However, as Pasley shows, the cheese actually had nothing to do with the mammoth; it was just a gift—albeit a strange one—and was nicknamed the "Mammoth Cheese" by someone else. The cheese was still knocking around the White House as late as 1804 and observed, by that time, to be "very far from being good."

3. Billy Possums

To learn about the effort to list the polar bear under the Endangered Species Act, I interviewed many of the people involved and slogged through many legal documents, including the original petition filed by Kassie Siegel and Brendan Cumming, "Petition to List the Polar Bear (*Ursus maritimus*) as a Threatened Species Under the Endangered Species Act." Kassie, in particular, was a patient and fair-minded explainer of the ins and outs of the case. I also benefited from conversations with Holly Doremus at the UC Berkeley School of Law, both about the polar bear case and the listing process in general.

The 2008 documentary *Polar Bear Fever*, produced by the Canadian Broadcasting Corporation, gives a good overview of the swelling of public interest in polar bears during that time. The 2007 UN report I refer to is

"The International Panel on Climate Change (IPCC) Fourth Assessment," and its lead author, Richard B. Alley, was quoted in the February 2, 2007, *New York Times* article "Panel Issues Bleak Report on Climate Change." Terry Macko at WWF discussed the *Golden Compass* campaign with me.

In researching the history of the Endangered Species Act and the candidate list, I read a number of good books, including Charles C. Mann and Mark L. Plummer, *Noah's Choice: The Future of Endangered Species* (New York: Knopf, 1995); Dale D. Goble, J. Michael Scott, and Frank W. Davis, eds., *The Endangered Species Act at Thirty, Vol. 1* (Washington, DC: Island Press, 2005); and two well-researched reports by the Center for Biological Diversity: 2004's "Extinction and the Endangered Species Act," by Kierán F. Suckling, Rhiwena Slack, and Brian Nowicki and 2005's "Progress or Extinction?: A Systematic Review of the U.S. Fish and Wildlife Service's Endangered Species Act Listing Program 1974–2004," by D. Noah Greenwald and Kierán F. Suckling. Two other important sources were John G. Sidle, "Arbitrary and Capricious Species Conservation," *Conservation Biology* 12 (1998) and Shannon Petersen, "Congress and Charismatic Megafauna: A Legislative History of the Endangered Species Act," *Environmental Law* 29 (1999). It's Petersen who notes that major newspapers devoted only one sentence to the passage of the law and who describes Congress as regarding it as "a largely symbolic effort." Brendan Cummings delivers his "Yes I voted to kill the polar bear" zinger in *Polar Bear Fever.*

Yale University's Stephen R. Kellert is considered a godfather of the emerging field of human-animal studies, and I relied on both conversations with him and a very large pile of his writings from the last forty-plus years to understand that research. The opinion poll about mountain lions and lousewarts, for example, comes from Kellert's "A Study of American Attitudes Toward Animals: A Report to the Fish and Wildlife Service of the United States Service of the United States Department of the Interior," published in 1967. My discussion of phylogenetic relatedness relies, in part, on his "Public Perceptions of Predators, Particularly the Wolf and the Coyote," *Biological Conservation* 31 (1985). And his book *The Value of Life: Biological Diversity and Human Society* (Washington, DC: Island Press, 1996) summarizes and elaborates on some of his most interesting findings. Michael J. Manfredo, at Colorado State University, was another good guide into this field. The study in which a particular animal is said to have been kicked "like a football" is "Human-to-Animal Similarity and Participant Mood Influence Punishment Recommendations for Animal Abusers," by Michael W. Allen et al., *Society and Animals* 10 (2002).

Findings mentioned in the footnote come from William Siemer et al., "Factors that Influence Concern About Human–Black Bear Interactions in Residential Settings," *Human Dimensions of Wildlife* 14 (2009); George Feldhamer, "Charismatic Mammalian Megafauna: Public Empathy and Marketing Strategy," *Journal of Popular Culture* 36 (2003); Lingling Xiang, "Animal Use in Award-Winning TV Commercials in China Versus the U.S." (masters thesis, University of Florida, 2008); Susan Clayton, John Fraser, and Carol Saunders, "Zoo Experiences: Conversations, Connections, and Concern for Animals," *Zoo Biology* 28 (2008); E. Paul Ashley, Amanda Kosloski, and Scott A. Petrie, "Incidence of Intentional Vehicle-Reptile Collisions," *Human Dimensions of Wildlife* 12 (2007); Jennifer Wolch and Jin Zhang, "Siren Songs: Gendered Discourse of Concern for Sea Creatures," in *A Companion to Feminist Geography*, eds. L. Nelson and J. Seager (London: Blackwell, 2005); R. J. Hoage, *Perceptions of Animals in American Culture* (Washington, DC: Smithsonian Institution Press, 1989); and Janis Wiley Driscoll, "Attitudes Toward Animals: Species Ratings," *Society and Animals* 3 (1995). John Fraser, a researcher with the Wildlife Conservation Society, told me about his discovery that people are more likely to presume a given tiger is female. This was in the course of his telling me many other, more important things.

I drew details about Roosevelt's bear-hunting trip from its coverage in the *New York Times* and *Washington Post* and from Douglas Brinkley, *The Wilderness Warrior: Theodore Roosevelt and the Crusade for America* (New York: HarperCollins, 2009), 431–45. My history of the teddy bear is rooted in a number of sources, all frustratingly incomplete. These include the Brinkley book; the Steiff company's Web site; Donna Varga, "Babes in the Woods:

Wilderness Aesthetics in Children's Stories and Toys, 1830–1915," *Society and Animals* 17 (2009); and Gary Cross, *Kids' Stuff: Toys and the Changing World of American Childhood* (Cambridge: Harvard University Press, 1997), 92–97. Cross describes earlier depictions of bears as "apparently designed to upset young children."

America's extermination of predators in the early 1900s is covered well in Barrow's *Nature's Ghosts* and in Lisa Mighetto's *Wild Animals and American Environmental Ethics* (Tucson: University of Arizona Press, 1991). I read about the *Ladies' Home Journal* story about Balser in Varga's "Babe in the Woods." The bureau biologist who claims predators "no longer have a place in our advancing civilization" is quoted in Stephen R. Kellert et al., "Human Culture and Large Carnivore Conservation in North America," *Conservation Biology* 10 (1996): 979.

I read, and actually kind of enjoyed, Seton's *The Biography of a Grizzly* (New York: Century Co., 1900) and parts of his books *Wild Animals I Have Known: Being the Personal Histories of Lobo, Silverspot, Rappylup, Bingo, The Springfield Fox, The Pacing Mustang, Wully and Redruff* (New York: Scribners, 1900) and *Animal Heroes: Being the Histories of a Cat, a Dog, a Pigeon, a Lynx, Two Wolves & a Reindeer and in Elucidation of the Same, Over 200 Drawings* (New York: Gosset & Dunlap, 1905). Seton's description of "shy" and "inoffensive" bears is quoted in "The Bear" by Daniel J. Gelo, in *American Wildlife in Symbol and Story*, eds. Angus K. Gillespie and Jay Mechling (Knoxville: University of Tennessee Press, 1987), 151. It was probably Gelo's essay that led me to consider Seton and teddy bears together.

Mighetto's *Wild Animals and American Environmental Ethics* puts the nature fakers in historical context, as does Ralph H. Lutts, *The Nature Fakers: Wildlife, Science and Sentiment* (Charlottesville: The University of Virginia Press, 2001). William Long's description of polite wolves appears in "The Sociology of a Wolf Pack," *Independent* 66 (1909). U.S. Census Bureau data tracks the urbanization of America during this time. The anguished zoo director I mention is William Temple Hornaday, quoted in Gregory J. Dehler, "An American Crusader: William Temple Hornaday and Wildlife Protection, 1840–1940" (PhD diss., Lehigh University, 2001): 151.

The cougar study I summarize is "Changing Attitudes Toward California Cougars," by Jennifer R. Wolch et al., *Society and Animals* 5 (1997). Montana governor Brian Schweitzer trash-talked wolves in a February 17, 2011, Reuters article. The study about New Jersey black bears is "The Black Bear Hunt in New Jersey: A Constructionist Analysis of an Intractable Conflict," by Dave Harker and Diane C. Bates, *Society and Animals* 15 (2007). I read only the parts of James Oliver Curwood's *The Grizzly King: A Romance of the Wild* (New York: Doubleday, 1916) that I absolutely had to.

I learned about the teddy bear's runaway popularity and the Billy Possum's rise and fall by reading many bizarre news articles of the time in the *Washington Post, Los Angeles Times, New York Times, Chicago Tribune,* and *San Francisco Chronicle,* among other papers. (I read, for example, that in 1909, a Mrs. John Rossman started breeding live opossums in her Brooklyn apartment and insisted to the *Washington Post* that the fact that fashionable women on the street were not yet carrying these Billy Possums around as accessories "is due entirely to the cold weather.") The "Christmas goose" line comes from Margaret Warner Morley, *The Carolina Mountains* (New York: Houghton Mifflin, 1913), 77. The Amazon.com review was posted by "Unusualfinds" on July 6, 2010. The reviewer goes on to say that she removed the toy opossum's hideous tail, took out some of the stuffing, shortened it, and sewed it back on: "Easy to do and made it much more toy-like looking, and less realistic."

My speculation about why the story of Roosevelt's bear hunt resonated with the public owes a lot to a conversation I had with Kierán Suckling, the executive director of the Center for Biological Diversity. We were discussing the more recent mass affection for polar bears but his thoughts on bears and humans in general stuck with me and resurfaced here.

The Obama administration's argument that the polar bear is a threatened, and not endangered, species is laid out most clearly—albeit not so clearly at all—in the memorandum "Supplemental Explanation for the Legal Basis of the Department's May 14, 2008 Determination of Threatened Status for Polar Bears," dated December 22, 2010. The federal judge's questioning of Kassie Siegel is quoted in "Judge Skeptical About Remanding Polar

Bear Case to Obama Administration," by Lawrence Hurly of *Greenwire*, published by the *New York Times* on February 23, 2011.

4. The Connection

I'm indebted to Brian Ladoon for his time and insights, and for protecting me from polar bears while I was at Mile 5. The *Canadian Geographic* article I mention is "Dangerous Liaisons," by Pauline Comeau, September–October 1997.

I learned to see Churchill from the perspective of wildlife photographers by talking with Daniel J. Cox, Mike Macri, and Norbert Rosing. Chris Palmer's *Shooting in the Wild* exposes the trickery going on in wildlife filmmaking. Also helpful was Gregg Mitman, *Reel Nature: America's Romance with Wildlife on Film* (Cambridge: Harvard University Press, 1999).

The radio show whose Web site revived Norbert Rosing's photographs of the bear and dog was called *Speaking of Faith* and is now called *On Being*. Stuart Brown, who assembled the package "Animals at Play" in the December 1994 issue of *National Geographic*, which included Rosing's photos, helped me reconstruct the chronology of events.

Margie Carroll's books, *Portia Polar Bear's Birthday Wish* and *A Busy Spring for Grandella the Gray Fox*, are published by the Margie Carroll Press: margiecarrollpress.com.

Thanks to the winners of 2010's Project Polar Bear contest for letting me and my family crash their grand-prize buggy ride, and a special thanks to Sam Leist for e-mailing me his video of the female yearling standoff so that I could better describe it.

5. The Lift

Daniel J. Cox's video of the starving cubs is, as of this writing, still posted on his Web site, naturalexposures.com. It is worth watching. Thanks to him for sharing the video with me and for discussing such a sensitive subject.

Information about Manitoba's Polar Bear Alert Program comes from Bob Windsor and Daryll Hedman at Conservation Manitoba and a video about the program produced by Polar Bears International.

The "Cold" episode of *The Martha Stewart Show* aired on the Hallmark Channel on December 6, 2010. It was quite good. The show is no longer on the air.

PART TWO: BUTTERFLIES

6. THE MIDDLE OF A HAIRCUT

While learning about the Lange's metalmark, and butterflies more generally, I was lucky to find a number of preternaturally patient teachers. I was also lucky that these men and women usually happened to be a lot of fun to spend time with. I'm especially grateful to Jana Johnson, Louis Terrazas, Jerry Powell, Richard Arnold, Travis Longcore, and Liam O'Brien. Liam's influence in particular extended far beyond the butterfly portion of the book, shaping my ideas about all conservation. His motto—"I just want to be part of a generation that tries"—may as well be the motto of this book.

Piecing together the history of Antioch Dunes, in this chapter and those later on, was not easy. Jerry Powell and Richard Arnold were incredibly helpful, in addition to being authorities on the Lange's itself, and both looked over sections of the manuscript, pointing out mistakes and pushing me toward more precise descriptions. I also benefited from conversations with the late Alice Howard, a champion for native plants throughout California, who worked at the dunes alongside Arnold, and Chris Nagano and David Kelly at the U.S. Fish and Wildlife Service.

Secondary sources I consulted include several revisions of Fish and Wildlife's recovery plan for the Lange's; the agency's most recent "5-Year Review: Summary and Evaluation" for the three endangered species at the dunes, dated June 2008; "Taking Refuge," by Matthew Bettelheim in the January 2005 issue of *Bay Nature Magazine*; Richard Arnold and Alice Howard, "The Antioch Dunes—Safe at Last?" *Fremontia* 8 (1980); Jerry Powell's unpublished study, "Changes in the Insect Fauna of a Deteriorating Riverine Sand Dune Community During 50 Years of Human Exploitation"; *Antioch* by the Antioch Historical Society (Mount Pleasant, SC: Arcadia Publishing,

2005); *Looking Back: Tales of Old Antioch and Other Places* by Earl Hohlmayer (Visalia: Jostens Printing and Publishing Division, 1991); and J. B. Roof, "In Memoriam: The Antioch Dunes," *The Four Seasons*, December 3 (1969).

Basic info about the Lange's metalmark, both here and in later chapters, comes largely from Richard A. Arnold and Jerry A. Powell, *"Apodemia mormo langei,"* in *Ecological Studies of Six Endangered Butterflies (Lepidoptera, Lycaenidae: Island Biogeography, Patch Dynamics, and the Design of Habitat Preserves)* (Berkeley: University of California Press, 1983), a book adapted from Arnold's dissertation.

7. SHIFTING BASELINES

Recently, Liam O'Brien built a Web site, sfbutterfly.com, a great resource for anyone who wants to learn more, and get excited about, the butterflies of the San Francisco Bay Area. I also learned about the region's butterflies from Arthur Shapiro at the University of California at Davis, and an undated and unpublished article Shapiro sent me called "Urban Survivors: San Francisco Butterflies Today," by H. V. Reinhard. The French lawyer I mention was named Pierre Joseph Michel Lorquin.

I learned more about Hans Hermann Behr, James Cottle, and the butterfly scene of turn-of-the-century San Francisco from stories in the *San Francisco Chronicle* and the *San Francisco Call*. These include "A Doctor's Career," *Morning Call,* October 1, 1893; "The Butterfly: Something About the Gaudy Ephemera," by Charles Belknap, *San Francisco Chronicle,* November 2, 1890; and "By Day He Catches Burglars; By Night He Catches Bugs," *San Francisco Sunday Call,* February 20, 1910. I also read Behr's essay "Changes in the Fauna and Flora of California," in the *Proceedings of the California Academy of Sciences* (1888) and "Butterflies—Try and Get Them," by Laurence Ilsley Hewes in the May 1936 issue of *National Geographic.* James Cottle's memoir is called "On the Wing—a Retrospect," and was published in *The Pan Pacific Entomologist* 4 (1928). Two histories by Robert Michael Pyle were also valuable: "Conservation of Lepidoptera in the United States," *Biological Conservation* 9 (1976) and "A History of Lepidoptera Conservation, with Special Reference to Its Remingtonian Debt," *Journal of the Lepidopterists' Society* 49 (1995).

Harry Lange recounted the day he caught Xerces in the Presidio in "Saying Goodbye," by Mark Jerome Walters in the December 1998 edition of *National Wildlife.* Thanks to Ed Ross, the entomologist who was with Lange that day, for meeting with me. I also learned about Lange from Hannah J. Burrack, an assistant professor at North Carolina State University, who'd interviewed many of his colleagues for a University of California, Davis, symposium in his honor, and from "Harry's Just Wild About Battling Bugs," by Art German in the *Sacramento Bee*, January 28, 1993.

The term "shifting baselines syndrome" originated in Daniel Pauly, "Anecdotes and the Shifting Baselines Syndrome of Fisheries," *TREE* 10 (1995). My other sources on the subject include Daniel Pauly et al., "Fishing Down Marine Food Webs," *Science* 279 (1998); and Karen A. Bjorndal and Alan B. Bolten, "From Ghosts to Key Species: Restoring Sea Turtle Populations to Fulfill their Ecological Roles," *Marine Turtle Newsletter* 100 (2003).

Peter J. Kahn Jr.'s writing on environmental generational amnesia is outright revelatory. See his book *Technological Nature: Adaptation and the Future of Human Life* (Cambridge: MIT Press, 2011), and "Children's Affiliations with Nature: Structure, Development, and the Problem of Environmental Generational Amnesia," in *Children and Nature: Psychological, Sociocultural and Evolutionary Investigations*, eds. Peter H. Kahn Jr. and Stephen R. Kellert (Cambridge: MIT Press, 2002), 93–116.

To learn about rewilding, I read Josh Donlan, "Re-wilding North America," *Nature* 436, August 18, 2005, and C. Josh Donlan et al., "Pleistocene Rewilding: An Optimistic Agenda for Twenty-First Century Conservation," *The American Naturalist* 168 (2006). The letters from the public—"colossal asshat" and so forth—are quoted in C. Josh Donlan and Harry W. Greene, "NLIMBY: No Lions in My Backyard," in *Restoration and History: The Search for a Usable Environmental Past*, ed. Marcus Hall (New York: Taylor & Francis, 2010), 293–305.

8. Our Vanishing Wildlife

Thanks to Dé Mackinnon, my mother-in-law, for sending me the newspaper clip about turtles causing trouble at JFK. I can't cite the specific article, however, because, not realizing its significance at the time, I threw it out.

I read about sea turtles in Columbus's time in Wilcove's *The Condor's Shadow*, 154, which relays the 660 million estimate. The subject is also covered in Bjorndal and Bolten's "From Ghosts to Key Species" paper. I read about bears ruining Internet connections in "For Idaho and the Internet, Life in the Slow Lane," by Katharine Q. Seelye, *New York Times*, September 13, 2011.

Other good accounts of early American wildlife can be found in Peter Matthiessen's *Wildlife in America* (New York: Viking, 1959) and Jennifer Price's essay collection *Flight Maps: Adventures with Nature in Modern America* (New York: Basic Books, 2000). Details about Martha's posthumous flight to San Diego come from newspaper stories about the trip and from James Dean, at the Smithsonian, who I also thank for a fun behind-the-scenes tour.

Details about buffalo in this section come from Dehler's "An American Crusader"; Barrow's *Nature's Ghosts*, 113–20, which does a good job of framing Hornaday's effort in the context of other conservation; and William Temple Hornaday's own book-length report, *The Extermination of the American Bison*, published by the Smithsonian Institution in 1889 and reprinted by the Smithsonian Press in 2002. The account of buffalo charging into a moving train is in Richard Irving Dodge, *The Plains of the Great West and Their Inhabitants* (New York City: G.P. Putnam's Sons, 1876), 121–22. The "rate of extermination" quote comes from "A Mighty Herd Has Gone," *Washington Post*, April 15, 1889.

Gregory Dehler's dissertation on Hornaday, "An American Crusader," was the key source for me as I tried to understand the man's life and work. In addition, I read Hornaday's *The Minds and Manners of Wild Animals: A Book of Personal Observations* (New York: Scribners, 1922); *Thirty Years War for Wild Life: Gains and Losses in the Thankless Task* (New York: Scribners, 1931); and *Our Vanishing Wildlife: Its Extermination and Preservation*, cited previously. I also drew from Hornaday's speech, "Last Call for Game Salvage," published in *Proceedings of the North American Wildlife Conference Called by President Franklin D. Roosevelt* (1936).

The truth is—if it's not already obvious—that I got a little obsessed with Hornaday. I spent many hours reading old newspaper stories about him, primarily in the *New York Times* and *Washington Post*. His complaints about litter at the zoo, for example, were published as "Director of Zoo Makes Protest," *New York Times,* May 28, 1908, and he remembered "Dohong," the philosophizing orangutan, in "Dr. William T. Hornaday, King Among Beasts, Tells of the Great Animals He Has Known," *Washington Post,* November 22, 1908. Hornaday's chart, in which the beaver scores 100 for "Original Thought," was published in *The Minds and Manners of Wild Animals,* 41.

The historian Frank Graham describes Hornaday as being written out of the history of the environmental movement in *Man's Dominion* (New York: M. Evans & Company, 1971), 207. Hornaday's unpublished memoir, which I quote from, is titled *Eighty Fascinating Years*, and part of the William Temple Hornaday Papers at the Wildlife Conservation Society, Bronx, New York. Details about his funeral were pulled from Dehler's dissertation and "Notables Attend Hornaday Rites," in the March 10, 1937, edition of the *New York Times.*

Much later, when I was done writing about Hornaday but still often found myself fishing through newspaper archives for stories about him anyway, I found one called "Dead Curator Calls Upon Live Ones to Preserve Bison Family He Slew," by Paul Sampson in the *Washington Post,* July 23, 1957. The article explains that the Smithsonian had recently dismantled Hornaday's taxidermy buffalo group—the one he assembled after his hunt in Montana. Workmen discovered a rusty metal box buried in the exhibit's fake ground. Inside the box was a note that Hornaday had written to his successors at the museum. "When I am dust and ashes I beg you to protect these specimens from deterioration and destruction," it said.

For estimates of the number of insect species, and their ecological contributions, I relied on Scott Hoffman Black and D. Mace Vaughan, "Endangered Insects," in Vincent H. Resh and Ring T. Cardé, eds., *The Encyclopedia of Insects*, volume 2 (San Diego: Academic Press, 2009), 320–24. To learn about the modern history of insect conservation—or the lack of insect conservation—I read "The Danger of Deception: Do Endangered Species Have a Chance?" Scott Hoffman Black's written testimony before the U.S. House of Representatives, Committee on Natural Resources Oversight Hearing, May 21, 2008, and the two historical studies, cited previously, by Robert Michael Pyle.

My discussion of the Hutcheson Memorial Forest and the assumption that "what nature needs most is for people to leave it alone" owes a great debt to Holly Doremus's brilliant paper "The Endangered Species Act: Static Law Meets Dynamic World," in the *Washington University Journal of Law and Policy* 32 (2010). Fortunately, I met Holly when I was just starting this book; it was exciting, and encouraging, to discover that she was asking a lot of the same questions and already had a few very compelling answers. I'm also grateful to her for lending a critical eye to several sections of the manuscript.

Additional information about the Hutcheson Memorial Forest is drawn from Daniel Botkin, "Adjusting Law to Nature's Discordant Harmonies," *Duke Environmental Law & Policy Forum* 7 (1996): 29–31; and "The Woods of Home," by Lincoln Barnett in the November 8, 1954, issue of *Life* magazine. Thanks also to Gordon Pratt, the lepidopterist who told me, "We can't just throw up a fence and think everything's going to go back to how it used to be"—for helping me better understand the butterfly's recovery.

To reconstruct the story of Humphrey the Whale, I spoke with Bernie Krause, who wrote about the rescue in his book *Into a Wild Sanctuary: A Life in Music and Natural Sound* (Berkeley: Heyday Books, 1998), 107–28; Jean Takekawa of the Fish and Wildlife Service; and Wendy Tokuda, a CBS television reporter who covered the rescue. I benefited especially from a long interview with Diana Reiss, who writes about Humphrey in *The Dolphin in the Mirror: Exploring Dolphin Minds and Saving Dolphin Lives* (New York: Houghton Mifflin, 2011), 1–22.

I also read coverage of the whale rescue in the *Sacramento Bee, San Fran-*

cisco Chronicle, Los Angeles Times, Associated Press, *Washington Post*, *USA Today, Newsweek*, and *New York Times*. Coverage by *ABC News* and *Nightline* was available on YouTube. *The Great Whale Rescue: An American Folk Epic* (New York: Pharos Books, 1986), by Tom Tiede with Jack Findleton, gives a valuable firsthand account. Findleton described his "sensitive feelings" in "Hooked on Rescue: Emotions of U.S. Rode With Whale," by Richard C. Paddock, *Los Angeles Times*, November 9, 1985.

The Fish and Wildlife Service detailed damage done by crowds to Antioch Dunes in the refuge's 1985 annual narrative.

9. WITHOUT CHANGE, THERE WOULD BE NO BUTTERFLIES

Thanks, again, to Jana Johnson at the Butterfly Project at Moorpark College, and to her students, for letting me throw questions at them while we all stared at butterflies.

For more about the effects of conserving top predators, see James A. Estes et al., "Trophic Downgrading of Planet Earth," *Science* 333 (2011). I read about the net worth of bats in a March 31, 2011, press release from the U.S. Geological Survey, "Bats Worth Billions to Agriculture: Pest-control Services at Risk," which summarized "Economic Importance of Bats in Agriculture," by Justin G. Boyles et al., in *Science* 332 (2011).

Robert Michael Pyle writes about the "extinction of experience" in *Thunder Tree: Lessons from an Urban Wildland* (Corvallis: Oregon State University Press, 2001), 140–53.

Information about Rudi Mattoni and the Palos Verdes blue comes from conversations with many people, especially, of course, Rudi himself. On the rediscovery of the Palos Verdes blue, also see "A Butterfly Flutters Back from the Brink," by Marla Cone in the *Los Angeles Times*, March 30, 1994. The Palos Verdes blue's initial, presumed, baseball-related demise is detailed in "Palos Verdes Blue Butterfly May Never Again Do Its Aerial Ballet," by Ann Johnson in the April 7, 1985, *Los Angeles Times*. Also useful were "Mister Butterfly," a profile of Rudi by Nick Green for the (Torrance, CA) *Daily Breeze*, September 27, 1999, and "The Importance of Farming Butterflies," by Ashley Morton in the May 20, 1982, issue of *New Scientist*.

I never had a chance to see Defense Fuel Support Point, San Pedro, for myself but based my description in large part on Travis Longcore and Catherine Rich's intriguing essay "Invertebrate Conservation at the Gates of Hell," in *Wings: Essays on Invertebrate Conservation,* published by the Xerces Society, Spring 2008. I gathered details about the military's role in conservation from the U.S. Department of the Interior's report, "The State of the Birds 2011: Report on Public Lands and Waters."

Jana wrote about her relationship to the Palos Verdes blue in the November 2007 issue of the inspirational magazine *Guideposts,* in an essay titled "Sanctuary." Mattoni's paper about the Sonoran blues at San Gabriel Wash is "An Unrecognized, Now Extinct, Los Angeles Area Butterfly (*Lycaenidae*)," *Journal of Research on the Lepidoptera* 4 (1989).

The city attorney of Colton, California, is quoted disparaging the Delhi Sands flower-loving fly as a "maggot" in "Rare Fly Buzzes Into Debate on Jobs," by Sandra Stokley, the (Riverside, CA) *Press-Enterprise,* May 12, 1997. Mattoni is quoted disparaging politicians in "Developers Wish Rare Fly Would Buzz Off," by William Booth, *Washington Post,* April 4, 1997.

Looking into the perceived, and actual, kinship between kids and animals, I read Kahn and Kellert's *Children and Nature*; Kellert's *Value of Life*; Edward O. Wilson, *Biophilia: The Human Bond with Other Species* (Cambridge: Harvard University Press, 1984); Gene Myers, *The Significance of Children and Animals: Social Development and Our Connections to Other Species*; and, most important, Gail F. Melson, *Why the Wild Things Are: Animals in the Lives of Children* (Cambridge: Harvard University Press, 2001).

Other details in this section are drawn from Alie H. Kidd and Robert M. Kidd, "Reactions of Infants and Toddlers to Live and Toy Animals," *Psychological Reports* 61 (1987), and Maarten H. Jacobs, "Why Do We Like or Dislike Animals?" *Human Dimensions of Wildlife* 14 (2009). It was the Kidds, moreover, who affirmed that children who don't care about animals are normal—a detail mentioned in Myers, *Children and Animals,* 177. (I am quoting Myers's paraphrasing of their conclusion.) The footnote on children, fear, and biophilia draws on Judith H. Heerwagen and Gordon H.

Orians, "The Ecological World of Children," in Kahn and Kellert's *Children and Nature*. Research about animal dreams is summarized by Paul Shepard in *The Others: How Animals Made Us Human* (Washington, DC: Island Press, 1996), 74–76.

The surveys of children by Stephen Kellert and Miriam Westervelt were published by the Government Printing Office in 1983 as *Children's Attitudes, Knowledge, and Behavior Toward Animals*. Kellert also describes this research in his book *Value of Life*, 44–47, which I quote from here as well.

The children's book *Humphrey the Lost Whale: A True Story* was written by Wendy Tokuda and Richard Hall and illustrated by Hanako Wakiyama (Torrance: Heian International Publishing Company, 1986). The *New York Times* article that lays out the advantages of teddy bears as compared to dolls is "The Baby Doll of All Nations," February 6, 1909.

10. THE SOUP STAGE

Information about Pokémon founder Satoshi Tajiri comes from what appears to be the only in-depth interview he's ever done: a Q&A with Tim Larimer for *Time Asia*, published on November 22, 1999. Thanks to J. C. Smith at the Pokémon Company International for answering some questions via e-mail, and to my nephew Sam Goldblat for making me curious about Pokémon in the first place.

The study of British children I mention is described in "Why Conservationists Should Heed Pokémon," by Andrew Balmford et al., in *Science* 295 (2002).

My writing about taxonomy owes a lot, again, to Holly Doremus and her work, particularly "The Endangered Species Act: Static Law Meets Dynamic World." In describing the ways infants perceive four-legged animals, I'm cribbing from Paul Shepard's *The Others*, 45–47. The evolutionary biologist Ernest Mayr proposed the "biological species concept" in *Systematics and* The Origin of Species (New York: Columbia University Press, 1942). Worry about "an undesirable trend toward taxonomic chaos" was expressed in J. Gordon Edwards, "A New Approach to Intraspecific Categories," *Sys-*

tematic Zoology 3 (1954): 2. The 1953 paper I quote from is E. O. Wilson and W. L. Brown Jr., "The Subspecies Concept and Its Taxonomic Application," *Systematic Zoology* 2 (1953).

The official description of Lange's metalmark comes from John A. Comstock, "A New Apodemia from California," *Bulletin of the Southern California Academy of Sciences* 37 (1938).

Thanks to Benjamin Proshek for sharing, explaining, and re-explaining his research on Mormon metalmarks, and to his academic advisor at the University of Alberta, Felix Sperling. Rudi Mattoni's Gulf fritillary experiment is written up in Thomas E. Dimock and Rudolf H. T. Mattoni, "Hidden Genetic Variation in *Agraulis vanillae incarnata* (Nymphalidae)," *The Journal of Research on the Lepidoptera* 25 (1986).

Thanks to Brent Plater, of the Wild Equity Institute, for several good conversations about his lawsuit. Also with us that day in Antioch was Peter Galvin, a founder of the Center for Biological Diversity; Galvin was assisting Plater with the Lange's metalmark suit and added to my understanding of what was at stake. The U.S. Fish and Wildlife letter I quote from was written by Cay C. Goude, an assistant field supervisor at the agency, to the California Energy Commission, August 17, 2010.

For an example of the right-wing response to Wild Equity's lawsuit see Jane Jamison's editorial, "Eco-Nuts Torture California Businesses with 'New' Pollutants," published on the Web site Right Wing News on January 22, 2011. Jamison warns that Californians will soon be paying more for electricity because of "nutty control-freak environmentalists who survive as parasites of government bureaucracies . . . in behalf [*sic*] of bugs which end up on windshields and car grills anyway."

PART THREE: BIRDS

11. CONSTRUCTION WORKERS

As I point out, the months I spent hanging around the men and women of the Whooping Crane Eastern Partnership happened to be an unusually fraught and uncertain period for the partnership—and especially for Operation Migration. I'm grateful to them for letting me in anyway. I've tried to portray some of the difficulties and disagreements I glimpsed because I believe that they reveal, rather than diminish, how admirable these people actually are.

Sincere thanks to John French, Brian Clauss, Barb Clauss, and Charlie Shafer at the Patuxent Wildlife Research Center; George Archibald, Joan Garland, Marianne Wellington, Eva Szyszkoski, and Barry Hartup at the International Crane Foundation; John Christian, Tom Stehn, Doug Staller, and Joel Trick, past and current U.S. Fish and Wildlife Service employees; and Marty Folk of the Florida Fish and Wildlife Conservation Commission.

Special thanks are due, of course, to the staff of Operation Migration: Liz Condie, Walter Sturgeon, Richard van Heuvelen, Jack Wrighter, John Cooper, Gerald Murphy, Geoff Tarbox, Trish Gallagher, Heather Ray, and Caleb Fairfax; and to Joe Duff and Brooke Pennypacker, both of whom inadvertently taught me quite a bit about life while teaching me about flying with birds.

Historic population sizes of whooping cranes are based on data from the Whooping Crane Recovery Team and Dr. Ken Jones, compiled by Betsy Didrickson of the International Crane Foundation.

The story of Josephine and Pete (and of Robert Porter Allen and George Douglass) is detailed in Faith McNulty, *The Whooping Crane: The Bird That Defies Extinction* (New York: E.P. Dutton & Co., 1966). To learn about this period, I also read Robin W. Doughty, *Return of the Whooping Crane* (Austin: University of Texas Press, 1989) and Cindi Barrett and Tom Stehn, "A Retrospective of Whooping Cranes in Captivity," *Proceedings of the North American Crane Workshop* 11 (2010). In *Nature's Ghosts*, Barrow also writes beautifully about Allen's work.

Louisianan Claude Eagleson's description of cranes circling like square dancers is quoted in Gay M. Gomez, "Whooping Cranes in Southwest Louisiana: History and Human Attitudes," *Proceedings of the North American Crane Workshop* 6 (1992): 21. "Two Nebraska Duck Hunters Kill the Last of the Pompous Bird," was published in the *Washington Post,* February 7, 1904.

The range and population size of whooping cranes before European settlement are described in the U.S. Fish and Wildlife Service's March 2007 revision of the International Recovery Plan (*Grus americana*), pages 9–10. The *New York Times* blamed the crane's extinction on its "lack of cooperation" in "Scarcest of Crane Vex Wildlife Service; Whoopers Dodge Efforts to Save Them," February 4, 1946.

Several documents helped me better understand WCEP and the transition it was undertaking in the fall of 2010: "The Whooping Crane Eastern Partnership Five Year Strategic Plan," December 2010; "The Whooping Crane Eastern Partnership Guidance and Partnership Transition Documents," August 27, 2010; and "The Whooping Crane Eastern Partnership External Review Program Review, Final Report," March 31, 2010, by Jane Austin, Leigh Frederickson, Dr. Devra Kleiman, Dr. Phil Miller, and Dr. Tanya Shenk.

Operation Migration's Web site was an invaluable resource in a number of different ways. I relied on it extensively for details about past migrations, for descriptions of flights I did not witness firsthand, or to check my descriptions of ones that I did. Also useful were issues of *INformation*, the magazine OM publishes for its supporters, and a DVD the group produced called *Hope Takes Wing*.

12. CRANIACS

Thanks to George Archibald for a great afternoon of talking and watching sandhill cranes in Baraboo. To learn more about his work with Tex, I relied on Doughty's *The Whooping Crane*, 105–6, and many newspaper and magazine stories, including "Dr. Archibald Dances with Cranes So Their Tribes May Increase," by Linda Witt, *People* magazine, April 24, 1978; "Man and a Bird Dance Together to Preserve Species," by Bayard Webster, *New York Times*, March 25, 1980; "Odd Couple's Mating Dance Finally Lays an Egg," *Los Angeles Times*, June 2, 1982; "What Gee Whiz Means," a *Washington Post* editorial from June 7, 1982; and "Peeping in the Shell," by Faith McNulty in the January 17, 1983, issue of the *New Yorker*. I also watched an excellent video about George and Tex produced by the International Crane Foundation.

Aldo Leopold wrote about cranes in his essay "A Marshland Elegy" in *A Sand County Almanac* (Oxford: Oxford University Press, 1989). When I visited George Archibald in Baraboo, he took me down the road to Leopold's old cabin and quoted a very long passage, from memory.

I interviewed William Lishman on Skype and drew additional details about his house and art from his Web site and from a February 20, 1992, article in the (Montreal) *Gazette*, "Underground Digs." Lishman chronicles his first flights with geese and the origins of Operation Migration in his memoir *Father Goose* (New York: Little Brown and Company, 1995).

Robert Horwich answered many questions about his work by phone and e-mail and pointed me to several helpful scientific papers, including Robert H. Horwich, John Wood, and Ray Anderson, "Release of Sandhill Crane Chicks Hand-Reared with Artificial Stimuli," *Proceedings of the North American Crane Workshop* (1998) and Robert H. Horwich, "Use of Surrogate Parental Models and Age Periods in a Successful Release of Hand-Reared Sandhill Cranes," *Zoo Biology* 8 (1989). A collection of studies detailing early attempts to teach cranes to migrate was published in *Proceedings of the North American Crane Workshop* 8 (2001). The remote-controlled whooping crane is mentioned in "Pilots Train Cranes to Fly Away Home," by Less Line in the December 9, 1997, edition of the *New York Times*.

Thanks to Mary Vethe and her third-graders at Pineview Elementary in Reedsburg, Wisconsin.

In the fall of 2008, I tagged along for a short stretch of Operation Migration's ultralight migration while working on a story about the project for the *New York Times Magazine* ("Rescue Flight," February 19, 2009). It was during that trip, and not during the 2010 migration described in the book, that I met Squire Babcock outside the church in Kentucky. However, that brief scene and a couple of stray quotes from WCEP members are the only material I've pulled into the book from my earlier reporting.

13. THEIR INCREDIBLE ESSENCE

Information on Humphrey the Humpback's return comes largely from the stories by the Associated Press, *Washington Post,* and *USA Today,* which headlined one of its articles: "Humphrey: Advanced, Dumb or Lost." The government spokesman hoped Humphrey wouldn't come back in "Humphrey Finally High-Tails It Back to the Sea," by Herb Michelson, *Sacramento Bee,* November 5, 1985. The woman wished she could talk to Humphrey in "Return of the Wrong Way Whale," by Cynthia Gorney, *Washington Post,* October 23, 1990. And the San Jose commuter voiced his strong suspicions that a supreme intellect was at work in "Humphrey Hysteria," *USA Today,* October 24, 1990.

The Internet videos I refer to are titled "How to Snuggle with an Elephant Seal" and "Touched by a Wild Mountain Gorilla."

I'm grateful to Joana Varawa, aka Joan McIntyre, for her time, forthrightness, and perspective on so many things, and for sharing sections of a memoir she's writing. I feel lucky to have gotten to know her. The book she edited, *Mind in the Waters: A Book to Celebrate the Consciousness of Whales and Dolphins* (New York: Scribners, 1974), is gorgeous and strange, and her first memoir, *The Delicate Art of Whale Watching* (San Francisco: Sierra Club Books, 1982), taught me much about her post–Project Jonah life. Thanks also to Eugenia McNaughton, a former Project Jonah employee, for meeting with me.

It was largely by reading D. Graham Burnett's leviathan-sized history, *The Sounding of the Whale: Science and Cetaceans in the Twentieth Century* (Chicago: University of Chicago Press, 2012) that I got up to speed on John Lilly's work, Margaret Howe's experiment with Peter, and the origin and significance of *Songs of the Humpback Whales.* Another good book, *Thousand Mile Song: Whale Music in a Sea of Sound* by David Rothernberg (New York: Basic Books, 2010), covers some of the same ground. I also read Lilly's two books from this era: *Mind of the Dolphin: A Nonhuman Intelligence* (Garden City: Doubleday & Company, 1967) and *Man and Dolphin* (New York: Pyramid Books, 1969). Donal Henahan encouraged *New York Times* readers to connect with their "mammalian past" in "Is the Art Song Really Out of Date?" December 13, 1970.

For details about the parade in Stockholm, see "What Happened at Stockholm" by R. Stephen Berry in the September 1972 edition of *Bulletin of the Atomic Scientists.* Joan McIntyre's most notable press clips include "Whale Love," by Nicholas von Hoffman, *Washington Post*, February 16, 1973; "Children Protest Whaling," by Donald P. Baker, *Washington Post*, June 3, 1973; "Watch Out, Hawaiian Whales, The Crusader Is on the Way," by Jean M. White, *Washington Post*, October 29, 1974; and "Stumping World to Save the Whale," by Harriet Stix, *Los Angeles Times*, November 20, 1974.

The story of Monique the Space Elk is covered by Etienne Benson in *Wired Wilderness* (Baltimore: Johns Hopkins University Press, 2010). Thanks to Etienne for sharing some of his leftover research as well.

To understand Greenpeace's standoff with the Russian whaling vessel, I relied, again, on Burnett's *The Sounding of the Whale*; on the 2012 documentary film *A Fierce Green Fire* by Mark Kitchell; and on coverage in the *New York Times*. I also interviewed the historian Frank Zelko, author of a forthcoming history of Greenpeace. (It was during that interview that Zelko compared *Mind in the Waters* to *Silent Spring*.) The herbicide problem at Antioch Dunes is described in an April 2, 2012, article in the *San Francisco Chronicle* by Peter Fimrite, "Weed Killers Threaten Lange's Metalmark Butterfly."

I learned about WCEP's tracking operation from Eva Szyszkoski and Joan Garland at the International Crane Foundation. John Christian explained why tracking the cranes was initially, legally, necessary. To think through the philosophical and legal complexities of wildlife tracking in general, I read yet another exceptional paper by Holly Doremus: "Restoring Endangered Species: The Importance of Being Wild," *Harvard Environmental Law Review* 23 (1999).

The situation surrounding bighorn sheep in Texas was first described to me by a pro-donkey activist there named Marjorie Farabee and later confirmed by Tom Harvey at the Texas Parks and Wildlife Department. Coyote-control tactics of the USDA's Wildlife Services division are described in

"Wildlife Services' Deadly Force Opens Pandora's Box of Environmental Problems," part of a stunning, three-part exposé about the agency by Tom Knudson in the *Sacramento Bee* that ran in 2012. Information about Cornell's Right Whale Listening Network comes from the project's Web site.

I first read about the Otter-Free Zone in Doremus's "Restoring Endangered Species: The Importance of Being Wild." I learned more by speaking with Steve Shimek, chief executive of the Otter Project, and even more from talking with Greg Sanders, whose final word on the subject is drawn from "Agency Seeks to Lift Otter Ban," by Sara Lin in the *Los Angeles Times*, October 6, 2005.

My description of early attempts to promote wildness in whooping cranes quotes David H. Ellis et al., "Lessons from the Motorized Migrations," *Proceedings of the North American Crane Workshop* 8 (2001): 143.

Joe Duff first explained to me that, once upon a time, Canada geese were cherished and rare. "They're beautiful creatures," he said. "They used to be harbingers of the changing seasons and legends of the fall. But show me one person who likes Canada geese now. And it's our fault!" I subsequently read about the history of the species in many archival newspaper stories, including "Road to Oblivion Runs Both Ways," an April 10, 1966, *Washington Post* column by then chairman of the Audubon Naturalist Society, Irston R. Barnes. For an overview, I consulted "Large Canada Geese in the Central Flyway: Management of Depredation, Nuisance and Human Health and Safety Issues," prepared for the Central Flyway Council by P. Joseph Gabig in 2000.

The U.S. Fish and Wildlife Service estimates that there are now 3.5 million resident Canada geese in the country. The Department of Agriculture is the federal agency responsible for controlling them. The total number of geese the government kills every year is hard to come by, but I got at least a rough idea from documents obtained through a Freedom of Information Act by Barbara Stagno of the animal rights organization In Defense of Animals. Discussing pigeons, I quote from "The Chattering Sparrow," *New York Times*, September 2, 1878, and "To Get President to Join in Pastime Is Big Hope of Many," *Washington Post*, July 26, 1933.

The growth of urban raccoon populations is documented in the PBS Nature film *Raccoon Nation*. "Our Most American Animal" was written by Polly Redford and published in the December 1963 issue of *Harper's*. It includes a photo of her young son, shirtless, cuddling with a raccoon.

14. SPOILER

My attempt to reconstruct what happened at Tooke Lake is based on conversations with Brooke Pennypacker, George Archibald, and two former trackers with the International Crane Foundation, Anna Fasoli and Sara Zimorski. I also e-mailed with Bev Hansen of the Hernando Audubon Society. Most of all, I'm grateful to Clarice Gibbs for her invitation to visit and for sharing her story.

The U.S. Fish and Wildlife Service's *2011 National Survey of Fishing, Hunting, and Wildlife-Associated Recreation: National Overview,* published in August 2012, states that fifty-three million Americans feed birds outside their homes.

15. BACKPACKS FULL OF ROCKS

Information about Jefferson County comes from the county's Web site.

In relaying Brooke's speech on the beach at St. Marks, I've also interwoven several statements he made to me months later when—still unable to shake what he'd said that afternoon at St. Marks, but also still incapable of understanding why exactly I was so moved—I told Brooke so and asked him to elaborate. This is the only place in the book where I've knowingly combined two conversations this way, or altered a person's quotes except for making minor adjustments for the sake of clarity or accuracy.

EPILOGUE:
THE MAN WHO CARRIED FISH

I'm grateful to Phil Pister for meeting with me at his home in Bishop. To get the story right, I also relied on Pister's essay, "Species in a Bucket," in the

January 1993 issue of *Natural History* and "Edwin Philip Pister, Preserving Native Fishes and Their Ecosystems: A Pilgrim's Progress, 1950–Present," an oral history done in 2009 by the Regional Oral History Office of the Bancroft Library, University of California, Berkeley. (The interviewer was Ann Lage.) Thanks also to Terry Russi for showing me around Fish Slough. John Muir wrote about the redemptive effects of trout fishing in the essay "The Animals of the Yosemite," in *Our National Parks* (New York: Houghton Mifflin, 1917), 231.

For more gruesome details about the farmer who snapped in anger at the pelican colony, see "Farmer Snapped in Anger at Pelican Colony," in the *Kansas City Star*, October 6, 2011.

Finally, a note about Rudi Mattoni. Not long ago, one of the journalists I hired to fact-check this manuscript e-mailed Mattoni and asked him, among other things, to confirm a few details about his insect-cataloging project on the Rio de la Plata. Mattoni responded: "I have left the area. I am abandoning the idea." He explained that he'd been discouraged to find the local community and the academic community to be unsupportive and completely uninterested in his objectives. "Another of life's small tragedies," Mattoni wrote. In other words, he had given up.

Except, he hadn't given up. Instead, he recently moved to New York City, he wrote, "where I plan to spend the rest of my days working on the art/science interface." He's now working with an artist to develop mass-rearing methods for butterflies, just as he'd once done for the government's moths, with the aim of building "living sculptures." That is, Mattoni has thrown himself into communicating the importance of biodiversity via the same, more imaginative channels that he'd tried to open with his art exhibit in Buenos Aires. We should all wish him luck.

INDEX

INDEX

INDEX

Printed in the United States
by Baker & Taylor Publisher Services